WW II

SPECIAL ASSIGNMENT

Short Stories of
Exciting
Headline Events
in the Life of a
SEABEE

Now Uncensored!
and Declassified!

WW II

SPECIAL ASSIGNMENT

Short Stories of
Exciting
Headline Events
in the Life of a
SEABEE

Clayton F. Rasmussen

Pleasant Word

110th N.C.B.

"SPECIAL ASSIGNMENT"

ISLAND "X"

Clayton F. Rasmussen

Printed in the United States of America

Packaged by Pleasant Word, PO Box 428, Enumclaw, WA 98022. The views expressed or implied in this work do not necessarily reflect those of Pleasant Word, a division of WinePress Publishing. Ultimate design, content, and editorial accuracy of this work are the responsibilities of the author.

ISBN 1-4141-0004-3
Library of Congress Catalog Card Number: 2003107685

Table of Contents

Foreword

Personally, I deny any claims of being an historian, and yet history is made up of people, events, and periods of time. These components of history are included in the following accounts, but they also include personal emotions of which most historical accounts are lacking. They are written for your enjoyment and entertainment, while being accurate to the best of my ability. Put yourself into the event as you read, and adventure along with me, enjoying the trip!

Having grown up in our beloved country, America, my loyalty and patriotism runs deep. Seeing the world drawing headlong into a world war, during my high-school days, we kept up with the news as then available by radio and newspapers. We realized nations were drawing up battle lines in diplomacy. America was quite strongly split between isolationism and global participation. We recognized two great potential enemies were Germany, and Japan. The isolationist forces desired our nation to remain independent from world conflict, even though the Administration desired to join European allies against Germany. The Japanese were a powerful force in the Pacific region, but as long as they stayed in their place across the ocean, America was content to remain at peace.

The citizenry was quite unaware of the constant, severe threat that Japan was considered by our military. However, the author's brother was in the Army Air Corps, based in Hawaii during the first half of the 1930's. Upon leaving the Army in 1935, he was placed on "call-up" status for the following four years, because the military expected an outbreak of war with Japan. At the same time the United States was shipping scrap metal to Japan. Often heard at the time was the saying: "Today we ship Japan scrap metal, tomorrow they will drop it back on our heads in the form of bombs."

Then came that awful day of infamy, December 7, 1941! Japan made a surprise, unprovoked attack on Pearl Harbor. Our Naval Fleet caught in the harbor was devastated.

War was declared then against Japan and their ally Germany. We had already heard of the atrocities done by the Japanese upon the Chinese, and now they were aiming at America. Our very existence as a nation was threatened. To say we were unprepared as a nation to go to war was a gross under-statement!

I was nineteen at the time, and found myself in a critical type of work; an engineering machinist, in a Scientific Laboratory involved in developing scientific instruments for critical need by both civilian and military customers.

By order of the government my job was frozen; I could not be drafted, nor could I enlist unless released by higher authority. My deferment was continued until I finally convinced Dr. Beckman, for whom I worked, to release me to go into the military. So on June 16, 1943 I was sworn into the Navy Seabees (CB's), on "Special Assignment."

The reader is alerted to the fact that the language used in the following stories, such as "Jap," or "Nip," used instead of "Japanese," was the jargon of that time. It was intended to convey the message of that day as we felt it, toward a savage, merciless enemy, who deserved no less disrespectful name. They were our sworn enemies. Kill them, or be killed ourselves; they gave us no other choice.

Having seen friends die, or being maimed for life by these brutal enemies, made a righteous anger well up, even in a Christian. War is terrible, but our nation was placed in a position of defending itself against enemies who designed to conquer the World, and subjugate their victims to brutalities unknown in our former days. During our early days in the military we were trained both physically, and prepared mentally to fight for total advantage over the enemies we would encounter.

As Seabees our specialty was to use our skilled craftsmanship to provide facilities for military operations, and to maintain them, all done in combat areas. So besides honing our technical skills we were also trained in combat techniques. We would be exposed in our line of duty to all the fierceness of the enemy's attacks. We must respond with equal or more effective tactics.

No historian can faithfully transmit the total truth about events during any period, or culture, without also knowing firsthand the emotions of that time period. America had been suddenly plunged into a war it did not want, and for which it was ill-prepared. Civilians suddenly became soldiers. The transition was traumatic, and affected lives in many adverse ways, as well as providing loyal disciplines. Individuals were taught to work together with others to provide a team effort that was for the common good.

The goal at the end of the tunnel was to win the war, so all-out effort was aimed at that objective. The enemy we were fighting in the Pacific was Japan. The enemy soldiers we faced were Japs, or Nips, for short, who believed it was glorious to die fighting. We believed in fighting to return home again after obtaining our objective of defeating our enemy. This difference in philosophy of life and death made for entirely different response mechanisms by the forces engaged in battle.

World War 2 was a period of national testing. It was to prove the moral fiber of a people. It would prove their commitment to one another, their cohesiveness, unity of purpose, and their belief in an Almighty God who raises up nations and puts others down, all according to His divine righteousness and eternal plan. It was a time for training me for the rigors of life in the future, to test my faith and trust in God, and to offer me the sweetheart who would be my marriage partner for life.

The military had full support from the citizenry. There was a little "conscientious objection" by individuals, but even as a Christian I would never have thought to object to serving my country in this time of crisis, even in the military. To be given all the blessings and benefits of being a citizen of this God-blessed nation without being available to defend it, was unthinkable. However, I do thank the Lord He never placed me in a situation where I had to directly take another person's life, but I do thank those who did it for us by necessity!

Now as you read these accounts of various events in which I was involved, realize you are being presented happenings from a period of time quite different from the time in which we live today. Do not be offended. I was a Christian then, and thankfully I'm still kept by my same Lord and Savior Jesus Christ today. His omnipotence is the measure of my expectations!

In providing research for the events recorded in these story accounts, I have relied heavily on personal memory. My mind was sharpened by emotional reaction to the excitement, or trauma of the events, as well as a desire at the time to write about them in the future. Chronology, and statistical details were gleaned from my personal diary kept in secret at the time, letters written to my sweetheart, which she kept in a perfumed, locked box, and letters to my parents. (Thankfully, they saved them).

Other details were recorded in our Battalion's official record book: "Contract Completed," which includes many pictures, and artist sketches, along with narration of the 110th CB Battalion activities over its entire activation period. The battalion was commissioned in the summer of 1943, and de-commissioned in the late fall of 1945.

My service days extended from June 1943 to February 1946. Other Seabees of our outfit have provided refreshment of memories, as we have met on occasion, and the Seabee Museum at Port Hueneme has offered documents that record actual declassified military information unavailable from other sources. The Public Libraries have revealed details of interactions of our various units during military operations.

This writing has certainly been a personal delight, and a spiritual blessing to see in retrospect the hand of God at work on behalf of one of His children. We are enabled to see His overall purpose for events, unknown at the time because of our direct involvement, which can now be seen in light of events which followed.

Please understand that the events and adventures recorded here were all a part of our total time commitment to perform the assigned tasks for which we were responsible. Our lives, 24 hours each day, were dedicated to our military service. What we chose to make of it was each person's privilege, and was in accordance with his character, interest inclination, and personality. In these accounts I believe you will detect what my motivations were.

(50 Years Later—Author Composes Memoirs)

Acknowledgments

I want to thank my three adult children, Cheryl, Gary, and David for their tolerance in their growing up years of my repeated relating of many bits and pieces of events from my experiences in WWII. Many times those stories were retold upon their insistence so their friends could hear. As adults they encouraged me to put these experiences in writing, so that others could also share them.

My dear wife, Cora encouraged me to start writing, as she put together the pictures I took during the war into albums which could be shared. She started this legacy by gathering our pictures taken over a period of four decades from storage boxes and organized them into photo albums which now fill an entire bookcase. This provided the incentive to write the history that accompanied those photos.

Cora also critiqued and edited my first story: "Divinely Directed Delays." She was also the guiding light of my life as I lived those dark days in WWII, later to be my marriage sweetheart. Having lived a most fulfilling life together, she has now departed this life for a far better place with our Lord God in His Heavenly Paradise, where we will meet again.

I owe much to my Seabee mates with whom I served, and who have shared refreshing memories which correlate with my own experience, through those often frightening events. Pictures and artistic sketches have been gleaned from our battalion history book, "Contract Completed," faithfully recorded, edited and published by several officers and enlisted men of our 110th Seabee Battalion.

The very fact that I'm still living today is entirely due to the gracious and loving care God gave me during the events which are hereby related. He was always my protector, friend, and my very salvation. He provided during those war years rewards for me of which I was often unaware at the time. He included a sweetheart who became my lifelong wife; a G.I. Bill which provided the education for my profession; lifelong friends; and above all my spiritual growth through it all.

(Lifelong friends, 110th Seabee Machine Shop
Crew on Tinian)

Cora, to you I dedicate these writings; you who offered such encouragement and love, by which you helped mold and shape me during those trying war years into that man you would eventually marry. Then we shared forty-one memorable years together, always deeply in love! A love that would touch many lives, and produce three lovely children who can now share these experiences with their spouses and their own children. I owe you my deepest love for all eternity, Cora! Love transcends all else, for true love comes from God Himself.

(Author, with Wife Cora, and Children David, Gary, & Cheryl)

Introduction

"Hey Kids! Meet Your Dad."

"To Know Me is to See Me in My Early Twenties!"

Take a Look!

Multitudes of military veterans of WWII could begin the memoirs to their children by a headline similar to the above. However, too few of them have documented their activities so that their offspring could know how their experiences helped shape their lives for the future. For those of you who have longed to know what went on in the everyday life of a serviceman in his training and in the combat areas, read along. In the privacy of your own home participate in the dangers, the excitement, the stresses, the adventures and the personal services rendered by such a young man. Although the pattern of warfare in the Pacific theater of operations, was quite different from that in the European theater, many similarities in personal trauma will be recognized.

The moral fiber of those participating was being molded and shaped. Perhaps that is why that generation has been sometimes referred to as the greatest generation. Their lives were forever changed, and would help them to shape the very makeup of our nation with its future accomplishments.

Documented here, for the first time, is a series of true-life, short-story accounts of special events in the life of a young enlisted man, in the U.S. Navy Seabees. Embroiled in the battling for Pacific Islands in WWII, he like millions of others was only one of the support troops for the Military. Yet each man had his own special adventurous, and often-traumatic experiences, and Clayton F. Rasmussen, Machinist Mate Shop, 3rd Class, relates some of the events in which he participated.

Except for the strictest censorship at the time concerning some of these events, they would have been headlines in newspapers across the nation. Yet they were only everyday happenings in the lives of those giving their all to make America again a safe place to raise their families. The success of these efforts, amidst the merciful protection of Almighty God, can be evidenced by the fact the children of this author are able to read these accounts in the safety of their own homes.

The Seabee battalion was an outfit designed to perform construction projects in the combat areas, or when necessitated, to join forces with the Marines for combat. The 110th Seabee Construction Battalion was especially prepared to build airbases, and in addition was chosen by providence to perform combat functions. Fortunately the author was selected to be a machinist in the Machine Shop, as well as being one of the first one hundred Seabees to see combat with the 22nd Marine Regiment in the invasion of the Eniwetok Atoll, in the Marshall Islands.

The next assignment for the Battalion was helping secure the island of Tinian in the Marianas, and development of the *largest airbase in the world.* Our final thrust was helping to bring the war to a successful conclusion, with delivery of the atomic bombs on Japan itself. Then to make a rounded career of Navy life, the last three months of his service was on the Aircraft Carrier Suwanne, CVE 27, touring to Okinawa.

Meet this young man!

(Clayton F. Rasmussen, Machinist Mate Shop, 3rd Class)

BEGINNINGS

PRE-MILITARY
INDUCTION
CAMP PEARY
GULFPORT
HUENEME

Chapter 1

The Training of a Seabee

Pre-Military

Upon graduation from High School, I had no knowledge that in a short while our nation would be in war, neither did I realize the work I was being trained and involved in, would lead me into military service. Eighty percent of each Seabee unit was made up of men with considerable experience in some particular trade or profession.

Following is a brief of my experience prior to being selected for Special Assignment in the Navy Seabees. This experience, along with special interest in precision and innovative machinist projects help provide background. From this pre-military information the reader may better understand the reason I was selected to perform some of the more creative projects in our Seabee unit, discussed in the following stories.

The basic training for Seabees was the same for all of us, but the pre-military experience of each tradesman varied infinitely. The Seabee unit was composed of a vast number of abilities, which led to our motto of "Seabees Can Do"! As a unit we could undertake *anything* and successfully complete it!

Thus the lessons learned from such synergistic cohesion led us to continue the practice into civilian life as we returned to our individual careers after the war. Therefore, this first chapter includes my particular pre-induction experience, followed by the special training every Seabee received.

Clayton Rasmussen—A VET

(Pre-military Preparation—Seabee symbol)

Pre-Military Preparation

WWII was a world event that none of us young people growing up in America during "The Depression" years expected. Hadn't Germany been defeated in 1918? Was not Japan such a different culture from our Western civilization that they would not pose a real threat to our nation, isolated from them as we were by the vast Pacific Ocean? Because our nation was so divided by views of Isolationism-Vs-One World perspectives, Germany posed no real threat to the U.S.; so we imagined. As we young people entered the work force in 1940 the employment picture was becoming better, after so many years of harsh and sparse living for those unable to find adequate employment.

Even as I ventured from Pennsylvania to California in search of work in the aviation industry, it was not easy for a young man of 17, just out of High School, to find employment other than part-time. So investment in the training of trade skills seemed most likely to result in a job. I enrolled in two aircraft sheet-metal night school courses.

Aviation ambition was an obsession. Having a very strong desire to be a pilot, it was inspiring to hear songs such as: "Men with Wings," and "Off we go, Into the Wild Blue Yonder." And to read about airmen Charles Lindbergh, Billy Mitchell, and air racing pilots such as Jimmy Doolittle, and Roscoe Turner. Then having the opportunity to personally meet Roscoe Turner at his hanger in Burbank was only to strike a stronger spark to the glowing embers of flying. This was capped off by living with my brother who had served in the Army Air Corps and was then working for Lockheed Aircraft Company.

Around the same time I met a man who was on his way to Canada to serve as a volunteer in the Canadian Expeditionary Aviation group on their way to England to help defend the British Isles from Hitler's threats. He offered to take me with him to enlist, but after considering the alternatives, I did not have the makings of a mercenary pilot. Another offer came to our night school class, providing opportunity for volunteers to train in the Ferry Pilot Program at a near-by military training field as a civilian flyer. But I turned that down for reason I wasn't ready for the military discipline, being we weren't at war.

(My First Aerobatic Flight was in this Ryan Sportster)

I was anxious to learn flying, and especially in the test pilot arena. I went to Van Nuys, in the San Fernando Valley, and convinced a pilot to take me on a trial flight in an aerobatic sport plane to see how I liked it. He took me up to altitude over the Griffith Park area and proceeded to do rolls, loops, spins, and sharply banked turns till I was ashen in face, with a squeamish stomach. Boy, I guess a fellow would have to get used to that type flying. I suffered all afternoon long with motion sickness. Maybe this flying game wasn't so much fun after all!

Having turned 18, I still did not have a full-time job, and the reason for this was bothering me and it wasn't paying my bills. At this time an evangelist came to the church we were attending, and I went to the meetings on the nights I had no class. One night he spoke on the subject: "Afraid to be different!" It was addressed especially to young people, and it really hit home with me. I had reached the point in my life when I must make my own decisions, not relying on what my parents wanted, because they were clear across the continent from me now. I must make my decisions on what the Lord desired for me.

I had accepted Jesus Christ as my personal Savior at age 13, so I knew I was a Christian, because I believed the Bible and what it said about being born again. Since then, and even before, I had been quite firmly grounded in the study of God's Word, the Bible. So when this evangelist described what I knew was my need, I made my *life* commitment to serve the Lord *wherever* He would lead me, even relinquishing my aviation aspirations if He so desired. From that point on I knew I was being led of the Lord and that He was opening doors for me, while sometimes closing doors where I was not to go. He was leading the way! What He wanted for me was what I wanted, which was to follow His will for my life.

One night shortly after that decision, a knock came on the door of my brother Marius' house where I was living. It was a man asking for me. He had observed my work on outside construction projects at the house, and had inquired of Marius about my schooling and discovered I had all the college preparatory courses in Mathematics and the Sciences. So he was there to ask if I would consider applying for work at Dr. Beckman's company in South Pasadena, "National Technical Laboratories" (NTL). Beckman was looking for young men he could train, and encourage to grow in the field of Science. He recommended me, and took me to their Laboratory the next day for an interview. I was hired, and became the 57th employee, including Dr. Beckman!

(National Technical Laboratories, NTL, 1940,
South Pasadena, CA)

On the day after Thanksgiving, while the rest of the employees had the day off, they offered I could come and start work. I appeared, with my supervisor, and was escorted to the rear of the plant to their apple orchard. Seeing a man in overalls on a ladder trimming trees, I was told to help him and haul the trimmings to a pile of brush to be burned. This I did for about two hours, and then, assuming the man doing the trimming was getting tired, I offered to relieve him on the ladder. He agreed, and instructed me on the method of pruning, and I climbed the ladder. We proceeded all day, alternating the trimming and hauling, and so went my first day of work at NTL. I was so pleased to have a steady, full time job!

Two weeks later, it was time to receive our pay-checks, and I anticipated this with joy. My friends said the President of the Company would deliver them personally, so I would then get to meet him. Here I was, the least experienced employee in the company and I would get to meet the President, and CEO. The time for check distribution came while we were eating our bag lunches, and as the big man in a business suit approached, I suddenly began to blush.

"Well, Clayton, now that you aren't trimming trees anymore, how are you doing on this more technical job we have you on?" Here was the man I had trimmed trees with and who I thought was the janitor! What a surprise, to say the least. That was my introduction to the man who probably had more influence on the career I would choose for my lifetime work than any other human being. I owe much to him, and appreciate the personal friendship of such a well rounded person. Friendly, extremely intelligent, a Physical Chemistry Professor, California Institute of Technology Trustee, philanthropist, entrepreneur, businessman and industrialist of the highest caliber, is only an introduction to this man's qualifications. At age 87 he was inducted into the Inventors Hall of Fame, and at age 89 he received the nations top science honor, the National Medal of Science, from President Bush, at a White House ceremony.

Even at age 100, he was still going into the office at the "Arnold and Mabel Beckman Foundation" to attend Board meetings, keeping up on the philanthropic projects that he and his wife Mabel founded. To date he has donated Hundreds of Millions of dollars to Medical and Scientific research projects.

After my first day of work at NTL, I started working in the chemical mixing facility which they called the "Green House" where we made a sterilizing chemical called "Sterimatic Solution." I don't know if the name for the facility came because of the color of the house, or if it was describing the chlorine haze that always hung in the atmosphere. Incidentally, the air we breathed there had a wonderful medicinal effect for preventing colds and flu.

After five weeks in that facility, Dr. Beckman himself commended me on my work and offered to promote me to the "case assembly" of their "pH Meters," which was a one man operation. I gladly accepted, and went to work in the lower floor of the main building. This responsibility grew to include inspection of the meter cases as they came from the wood shop and lacquer finishing facility. After about three months I was approached by the supervisor of the Machine Shop, who had been looking over my personnel files and discovered I had taken courses in Sheet-Metal. He was about to start a sheet-metal facility as part of the Machine Shop and wondered if I would be interested in being the man to set it up. Always on the lookout for further opportunity, I heartily agreed. Then came a time of waiting for this to happen.

After about six weeks I was getting quite anxious, so I approached him to see if he had given the job to someone else. They had been delayed on the sheet metal equipment, but said he was expecting a different machine to arrive, any day soon. It was a new "Hardinge" Screw Machine, which no one in the shop knew how to operate, and he wanted to know if I was interested in becoming a machinist, specializing in screw machine operation and set-up. I jumped at the chance to learn something new!

He notified my supervisor that I was to be transferred to the Machine Shop the following week. Learning the *operation* of the screw machine was fascinating to me and after awhile he started to teach me the *setting up* of the jobs on the machine. The multiple operation screw machine performs nine or more different operations on a part that is in the lathe to be machined. To set-up the operations requires being able to precision grind cutting tools of various shapes, setting them in the turret in the sequence determined for the machining operations to take place, and adjusting the various feed speeds.

(Hardinge Screw-Machine I operated and Set-up)

This also gave me some training I would need to operate various other machines, such as lathes, milling machines, shapers, etc. Then came the opportunity to train another man on operation of the screw machine so the company could have us operating it on a 24 hour basis. I would set-up the machine, which usually required about six hours, and then turn out production parts for the rest of my 12 hour shift. The other man would produce the parts during his entire shift. The fun part was the setting up, which I alone was trained to do. Then came my training on other machines, with the lathe being my favored speciality.

A sheet-metal facility was also set up as my responsibility, and I trained men on the manufacture of our various sheet metal products, which included lay-out, shearing, bending, special fabrication, punch press, and spot-welding. I was able to fabricate some of the prototype sheet metal parts for new instruments being designed.

The Engineering Department had a prototype shop where they did their own machining, but when hearing of my work they asked that I be made available for specialty work for them. So I became their lab machinist technician. Every job was something special; for feasibility studies, or to determine if it could be machined on equipment we had available, rather than send it out to specialty shops. These parts really required ingenuity, and helped me develop a design sense, often suggesting changes in design to simplify the manufacture.

The Chief Designer/Draftsman, who later became the Chief Engineer, and I often planned together and worked out the designs. One night while I was working on some specialty Spectrophotometer "scrolls" which control the spectral scan of the instrument, I had a visitor, Dr. Beckman. He was working on a top-secret project for the Government, and needed to personally machine some parts in an adjacent lathe in the lab. Asking for my assistance in setting up the job, he proceeded to make the parts, and to keep me company for the rest of my shift. In the process, he promised that in the future he would have me make the parts for him which he would be designing from time to time. That was the beginning of my participation in the "Manhattan Project," even though at the time I had no idea this work was a link in the development of Radar and the Atomic bomb.

Specialty projects became my forte'. I enjoyed this part of the work to the fullest, as it enabled me to innovate, design, and to use my God given expertise to develop instrument parts that others found too difficult. Little did I realize that this same inventive gift was to follow me throughout my entire career and lifelong endeavors.

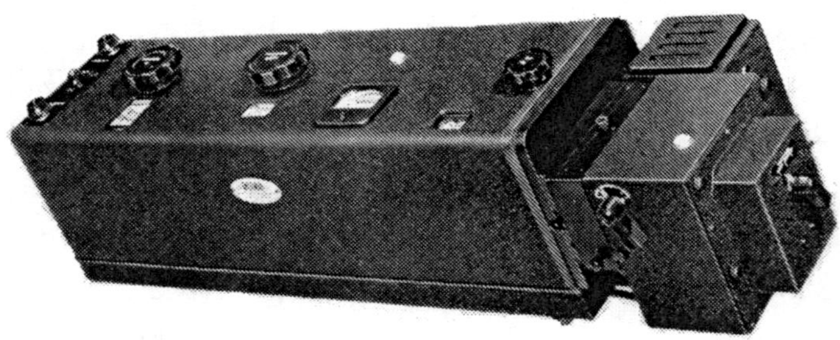

(DU Spectrophotometer—A Scientific Breakthrough!)

One of the problem parts that was needed to adjust the mirror for the spectrum scanning scroll, on the Spectrophotometer, was an adjustment screw that required the finest screw threads possible. Our lathe was designed to make 192 threads per inch. No facility however was found that could manufacture such precision threads in the quantities we required. So Engineering asked me to try my hand at producing one. The precision and sharpness of the thread-cutting tool was something beyond anything we had ever done. Using a magnifying glass, to be able to see the fine threads, I was able to machine the threaded bolt, and then with more innovation to bore a thread on the inside of the part into which it screwed . It was successful, and a precise procedure was established where I was able to machine these in quantity. They certified me as the only one to do this machining. Later, I was able to train another man to make the part.

As the war was taking most of the young men into the military I saw a need for a machining device to enable new, inexperienced personnel to turn out precision parts on a lathe. So I set out privately to design a "Lathe Cross-feed Turret Stop" on my outside spare time. The Manager gave me permission to use the company's machines to fabricate the parts, all on my own time. Upon completion of the device I had a newly employed young lady fabricate some very precise production parts utilizing my device, and all the parts passed inspection.

(Lathe Cross-feed Turret Stop)

The management was pleased, and offered to purchase the first one I would have produced, and the patent would be my own, since it was all my own idea and labor. In the next several weeks I had a patent search made by an attorney, blueprints drawn, and was prepared to search for a company to manufacture it for me. However, before being able to complete the process I was called into the military, and the project never went to the production stage. All I have to show for it is the prototype I built.

The war was churning on, and our forces were starting to gain a little on the offensive in the Pacific. The Newsreel Theater featured pictures of the military campaigns being waged in the European and Pacific theaters of war, and I was quite moved patriotically to be a part of the real fighting forces. Many of the young men in the company were being drafted into the military, but Dr. Beckman had gotten me deferments every six months, which froze me to my job, where the draft board could not take me. Very few young men were left as civilians, and I did not realize how important the instruments were to the military, that we were producing. So I requested to be released from my deferment, to join the Military, and received Dr. Beckman's reluctant approval.

I felt very strongly needed to do more to help defend our beloved country in this time of national crisis, and being on the firing line was my choice. In anticipation of immediate drafting into the military, I arranged to take a two weeks trip back to my Pennsylvania stamping grounds, to bid farewell to my friends there.

(Christmas in Centermoreland, PA, 1942)

After having a wonderful time actively participating in winter sports with many of my friends, and much visiting around town, I also arranged correspondence pals with some of my young lady friends. I was staying with a good friend Charlie Dempski, and on Christmas Eve we drove to the city for a night of roller-skating after which we arose next morning with a most beautiful "White Christmas"! We celebrated a wonderful down-on-the-farm dinner after opening presents.

Meanwhile I had received a letter forwarded from my parents in California which notified me to report to my Draft Board on January 5, 1943. This was the letter I had been expecting, so it was time to return to the West Coast and home.

(Dempski, Harding, Weiss, Rasmussen, Statnick)

Each of my Pennsylvania buddies pictured above, participated in the war effort in various ways. Dempski—Army Air Force; Harding—Navy Research Labs Engineer; Weiss—Army, (killed in action); Rasmussen—Navy Seabees; Statnick—Army Electronics Technician.

Taking the train to Chicago, I arrived at the station to catch my West-bound train, the "El Capitan." However, though I had a formerly confirmed reservation, they had no record of it. They had instead, loaded the train with military troops who took priority over all other personnel. The first train I could schedule out was ten days later. Thus, delayed in Chicago, I was unable to make it to my Draft Board appointment by the scheduled date. I sent a telegram to Dr. Beckman, asking him to inform the Draft Board and explain my inadvertent situation. Those ten days in waiting were spent visiting my brother Alvin and family in Evansville, Indiana.

On the train traveling home, every time I went to the dining car, or walked around, I was taunted by servicemen who loudly exclaimed: "Mother with a baby coming through!" Or: "Make way for a 4-F'r!" I was such a rarity, being of draft age and yet not in uniform, and the insults hurt. When I finally arrived home, I found that Dr. Beckman had gone to the Draft Board and obtained another six-month deferment for me!

Well, now I was back at the Lab, and awaiting my new assignments. They brought me a book of blueprints that was about 3-inches thick. It was all the prints of parts for a new instrument for which the Government had contracted with Dr. Beckman; the very first commercial "Infrared Spectrophotometer." It was later to receive the designation, the Beckman IR-1. I was to make three of every part, to be assembled as three of these instruments. So I was given the authorization to use any of the machines, with priority over any other job, and to complete the project in utmost urgency.

One of these instruments was to be put into the care of a special courier upon completion, to be delivered for use of the Federal Government. This was another part of the Manhattan Project, still totally unknown to me. The second one was to be used in the manufacture of TNT, and the third was for the project of making synthetic rubber, since Japan had confiscated our source of natural rubber. The instrument was also used to develop the mass production of Penicillin. This job took me three months.

Upon completion of the parts, the General Manager and I met at nights and weekends to assemble them. Then he took them to be tested. We packaged one for shipping and on a Sunday afternoon it was picked up by a special courier and he left on an airliner, having the instrument with him at all times. He was instructed to get it to its destination at any cost! Where that destination was, we did not know at the time, but found out later.

Military Induction

Upon completion of my six-month deferment, the Draft Board of Highland Park, Los Angeles lost no time in sending my invitation to join the military forces of Uncle Sam. I was classified 1A, which indicated immediate draft! I packed my things at the home where I roomed and boarded, and went to my parent's house in San Dimas, where I conducted all the last minute preparations before leaving for the Service. I had no particular choice for branch of military, except I hoped to get into the Army so I could train to fly in the Air Force. I heard from the friends at the rooming house that the Navy wasn't taking anyone, no matter how qualified they were. That suited me fine, because the Navy was the last place I wanted to be, as I was very prone to motion sickness, and being on a ship would be a nightmare.

June 16, 1943 was recorded on the calendar the morning I boarded the bus, for my appointment with the military. This was a day filled with activity. Arriving at the Court House in downtown Los Angeles, I followed the signs for new inductees, and ended in a line for our physical examination. This was at the time many of the Mexicans were Zuit-Suiters, with their black baggy pants and dress jackets, with their twirling gold chains hanging the length of their pants, and a fancy black hat with trim. Many of them belonged to the Pachuco gangs, and their female gangs were known as the Black Widows. One of my best friends at NTL was a Mexican, and he had warned me of their activities, and to be very careful around them, for most carried knives!

In this line of inductees were several that fit their description, and they were more brazen, loud and outspoken than anyone I had ever seen. When they were accosted by the Navy Corpsmen who where giving the orders to undress, it was a dangerous comedy. We all expected a fight would break out, by the disgraceful way the corpsmen were maligning them. These two parties were living up to the way "Life Magazine" had been portraying pictures of the fights between Zuit-Suiters and Navy personnel.

We were given orders to strip naked, and stuff our cloths in bags provided. For most young men this was a real test of their modesty. Standing stark naked in front of a bunch of strange men, while the worst was yet to unfold. First we were examined minutely trying to find scars by which we could be identified, in case we might be found lying on the battlefield, or stuffed in a body-bag. My only scars were four on one knee where I had been gouged when accidentally flipped from my coaster wagon as a kid, and one on my elbow received when falling off a trapeze.

This was quite different from the knife scars they found on some of the Mexicans, and scars from operations on others. These were recorded on one of our medical cards. Then we proceeded through a maze of curtained booths and aisles, getting checked for blood pressure, heart rhythms, lung soundings, temperature, hearing, eyesight, flat-feet, dental, genitals, rectal, and you name it as everything that makes a person tick. Rounding one curtain I was embarrassed to see the next doctor to examine me was a woman. That over, we soon discovered that several others of the doctors were females, who seemed very matter of fact to see all these men in the nude. Finally to the end, the last doctor signed us off as either fit for service; 1-A (draftable), or 4-F (medically disqualified). I had no problems, so was passed on to the next procedure.

We then dressed, and reported with our papers to a screening board, of three men in their respective uniforms; Army, Navy, and Marines. Here I had my choice, and handed the papers to the Army guy, (hoping for the Air Force). He looked at them briefly, then handed them to the Marine, who made some notes on them, and gave them to the Navy man. After looking in detail, and making more notes, he handed them back in a large envelop and told me to report to room 222 on the second floor. I proceeded to the referenced room and saw the word NAVY in big letters on the door. There had evidently been some mistake and I was given the wrong room number. So running downstairs again I approached the Navy man and told him of the mistake. He very tersely told me: "We don't make mistakes here, we correct them! You had the right room."

Well, I didn't return up the stairs as fast as coming down, and was thinking many things. *We* really don't know what's going on; and *they* sure don't know what they're doing! Inside this next room I found another line, and no one seemed to know any more than I did about what was happening. Finally my turn came to be interviewed by the first Navy Officer behind closed doors. We talked, as he looked over the papers, and remarked about my industrial experience, and my amazing deferments, at an advanced age just short of 21 years. Most of these incoming men were 17, or 18 years. It seemed like a good, friendly interview, and he concluded by stamping in big red letters the cryptic words: "SPECIAL ASSIGNMENT!"

Back in the next line I tried to find what this special stamp meant, but no one knew. They questioned: "Are you an FBI agent or something in Intelligence?" I sincerely hoped it didn't mean I was to be a spy, or something like that. No one was able to shed any light, except that one of the other men had the same stamp on his papers, and he was as ignorant as myself about it. At the next interviewing officer, he disclosed I was being assigned to the Navy Seabees. I still was as ignorant as before, as to what they were.

Next stage they told me was to be sworn into the Navy Seabees in the United States Naval Reserves, (USNR). The swearing-in ceremony would be later in the day, on the main floor of the Court House.

About thirty of us lined up and swore our allegiance to the U.S. and to serve our country even to the death, if required. Quite significantly as you will later learn, standing next to me was a man by the name of Red Mack. We were then given the surprising good news; we were being given a seven-day leave, and were then to report back at the Union Station in Los Angeles, ready to board the train for parts unknown.

On the way home, I was mulling over this surprise situation, being a member of the U.S. Navy! What a revolting development that was. I had made all my plans for flying in the wild blue yonder in a fighter plane, and now to be stuck on a ship was almost unbelievable. I thought the training would most likely be in San Diego, so at least it was not that far away from home. When my folks heard, they were about as surprised as I was.

The big question on our minds: what are the Seabees? So I made a quick trip to the library to read as much material as was available, which I found was only a couple of magazine articles, no books. This outfit had just been formed one year before, and it was composed mostly of demolition men, whose greatest exploits so far had been as the Navy Frogmen, or Under-water Demolition Teams (UDT). They had made quite a name for themselves in the first invasion off Tulagi Island just before the battle for Guadalcanal. They had landed on the beach by submarine, under cover of darkness, and destroyed all enemy communications before the main landing force of Marines hit the beach. They left a sign on the beach, (pictured in the article), saying: "Welcome, Marines—signed Seabees!"

My heart sank! What would I tell my parents, when they asked what kind of an outfit I was in? I told them the Seabees are an outfit that is called: Construction Battalion, the Seabee nick-name coming from the first letters of their functional name (CB). They also sometimes are called to go into the combat areas with the Marines; to construct facilities.

The remainder of my leave was occupied in revisiting my friends at NTL, and at the boarding house. They found it difficult believing I was in the Navy, since some of them had been refused when they tried to enlist, and now I was telling them I was drafted into the Navy!

I signed my car, a 1930 Model A Ford Coupe over to my Dad, since he had no car. I wouldn't need it for the duration of the war anyway. They gave me a handy, thin Bible to use while in service, and their church gave me a sendoff, and promised to keep me in prayer. I was also listed as a serviceman on the large plaque at the Church of the Open Door, where I was a member. The city of San Dimas put my name on their "active servicemen" display board in the park.

My parents and I had a good prayer meeting together as they committed me into the Lord's keeping power. I confided to them that I had asked the Lord to bring me home after the war, entirely whole in body. However, if it was His will that something would harm me, to allow me to be killed rather than be maimed for life. Yet I trusted He would see me safely through it all, and would have a lifelong purpose for me in His service. My life was entirely committed to the Lord's purpose for me, wherever it led, and through any circumstances.

(My Model-A Ford in Front of My Parent's Home)

Camp Peary—Cross Country

The day arrived for me to meet my new Navy mates at the Union Station for our train trip to *somewhere*. This was just a little taste of the faith and trust we had to place in the military authorities to see that they got us to where we were supposed to go. Mother joined me on the bus to Los Angeles, as she was bound for her son Marius and family, in Burbank, to visit with them for a few days while Dad was away. This was also indication of the sensitivity Marius and Ruth had for Mother, knowing she would be very lonesome now with her youngest son leaving for the Navy.

For me this was a time of excitement and anticipation! I didn't realize the trauma this separation was for Mother. It dawned on me when departing the bus in downtown Los Angeles, when I saw how bravely she was in trying to hold back the tears. We embraced and kissed in parting, and each went our separate ways. I joined a large crowd of new inductees, at the Union Station, all headed for the same place we were, all still dressed in civilian clothes. Uniforms would not be issued until we arrived at camp.

Several military officials made brief speeches, and provided instructions. Then they escorted us into a large hall where tables were set, and we enjoyed our first free banquet! It was delicious, and plenteous, as they made it known this would be our last formal meal for four days while we traveled to Camp Peary, Virginia. We would eat box lunches during that time. This burst my first bubble to think we would be clear across the country, instead of in Southern California where other Navy sailors were trained. I didn't know that our training was to be something different than the Regular Navy men, but we would soon find out. It was all quite exciting!

Soon the order: "Move em out!" was heard, and we filed to the particular train cars to which we had been assigned. We selected the bunk which we would inhabit for the next four days, and laid claim by dropping our bag of clothes there. Red Mack was in the bunk adjacent to mine. This was a time of introductions, and I met Horace Trigg who seemed to be a preacher kind, Bill Dally who was a giant of a young man, and Jack Griffin who was also a buddy of Bill's. None of us knew what rank each had come in as; we were just a bunch of young men headed into the great unknown. Until we put on uniforms with our designated ratings we were just friends of equality.

The train on which we headed East was a large coal-burning steam engine, not at all like the diesel engines and speed of the "El Capitan" which I had so recently ridden. Our car was full of double-deck bunks, side by side, length wise across the width, with one aisle along one side. We had no dining car.

There wasn't much to do except lie on our bunks, or visit. Trigg came through the car announcing we would be having a Sunday morning church service the next day. He would be the leader. Red and I lay on our bunks and shared accounts of our families, and what we expected out of our time in service, and our backgrounds. We were delivered box lunches each meal time. This was a fairly slow train, so we realized why it would be days before reaching our camp.

The first test of patience came when the train went through the first tunnel. Everyone had opened their windows for fresh air, and all at once as we entered the tunnel, it became dark, and soot, cinder-sand, and black, sulfury smoke filled the car. What a mess! The cinders were in our bedding, and the smoke clogged our lungs with sulfur fumes and black tar, leaving a coughing, choking bunch of gripping men. It was too late to close the windows, and besides, some of the windows now wouldn't close. We had learned our lesson! At night we slept in our cindered sheets which punctured our skin, setting up rashes and itching. Boy, what fun? We had a lot to learn, that we weren't in this man's outfit for comfort!

At Albuquerque we were allowed to get out of the train and buy snacks. We were afraid to go very far from our train for fear we would be left. I thought; what if we got left, where would we go; we had no idea how to get to the camp, and we would be AWOL even before we got into uniform. I stayed close to the train. Already I was learning that someone besides myself was taking care of me, and we were to just do as they said and ask no questions. Don't even think for yourself, because you may think in a different direction than where they wanted you to go. This was something really different for me, one who liked independence, and to do my own thinking.

Miles went by, and one morning we pulled into the rail yards of Chicago. We spent most of the day there, being shuffled from one track to another, with different engines pulling or pushing us, and seeing nothing but a lot of tracks and trains. This was the junction point where we were to take the train to the South, but I think they couldn't make up their minds which track went South, or at least that's what it seemed. Maybe it was to confuse the Fifth-

Columnists who were spying on troop movements, we conjectured as we passed the time; very much bored and anxious to get to camp.

Finally in late afternoon they must have found the correct track for we pulled out in a single direction, heading away from the setting sun. We slept quite well as we anticipated arriving at Camp Peary sometime the next day if all went without hitch.

Camp Peary—"You'll be Sorry!"

Quite exhausted, and worn out, we arrived in Williamsburg, Virginia our destination. We gathered our dirty clothes, and boarded Navy trucks and headed out of town, and it seemed we hit every bump in the road. We pulled up to a military gate, and after some formalities, the busses were allowed through. It seemed the guard didn't like our looks. All along the road we would see skinheads with clean white caps yelling: "You'll be sorry!" in a sing-song fashion. We wandered through a maze of barracks until we arrived at a section reserved for first-timers. We chose our bunks, which were double-deck and looked like they were made for very skinny guys. Then a Navy Chief gave us instructions, and vocabulary as to what to call the ordinary facilities. There was the "head" (bathroom), "deck" (floor), "bulkhead" (wall) "chow" (eats), "galley" (kitchen). Then he described the "short-Arm inspection" (a very intimate inspection of your person looking for venereal disease), "skullery" (pots & pans to be cleaned), "scuttlebutt" (rumors, or gossip), and other nonsense vocabulary. We all wondered why they didn't use the ordinary words with which we were familiar. We would sleep on used pads until we received our personal issue the next day.

All this was enough for one day, so we went to chow, which consisted of a line seemingly a mile long, ending in the dining hall. After chow, we retired to the barracks, and claimed our bunks. That night was a nightmare! Kids who had never been away from home suddenly realized where they were and cried all night. One case not far from my bunk, suddenly erupted in a scream from the guy in the lower bunk as he was treated to a dousing by the one in the top bunk who wet his bed in nervousness! Another had a nightmare and started beating up on the man in the top bunk, thinking he was being attacked by an eagle.

The "kind-hearted" Chief was able to settle all his children back into their bunks after each episode. More than one man fell off the top bunk, suddenly awakening as he hit the floor. I mentioned the bunks were only narrow enough for skinny guys. We would all get used to things in a few days; but it would take some training, which the Navy was anxious to get underway!

(Wake-up Call)

Until going through something like this Boot Camp you can't realize how complicated it is. Up in the morning at 5:30, calisthenics, followed by personal chores, breakfast after waiting in a long line, back to barracks to prepare for inspection of beds, and a briefing of what the day was to hold in store.

("You'll be sorry")

First on the list was another complete physical exam. Then everyone went to the barber to get his head shaved. When my hair is cut short it just plasters down flat on my head; absolutely nothing attractive.

Everywhere we went we heard the sing-song: "You'll be sorry!" Oh, how glad we would be when we had navy uniforms so we could be the ones singing this. Well the time came to go through the fitting for skivvies (underwear), chambray shirts, dungarees (pants), p-coats, caps, boots, socks, ponchos, rain-gear, and then came dress whites, dress blues, and dress shoes. Other things would be added as they were needed.

(Learning to Scrub Clothes)

Next they issued us the needs to do our laundry. Buckets, bar soap, and scrub brush. Not knowing how to use these utensils they trained us how to lay our pants on a scrub board and throw water on them, then rub with bar soap, and scrub as hard as you can with the brush. It's amazing how this can get them clean even after the muddiest days. Of course, our knuckles were worn thin. How you dry them and press them is a trick you learn by putting them under your mattress and sleeping on them. Even dress whites can come quite pressed this way. We learned by watching others goof, that you don't put your pants in a bucket of water and then pour bleach over them. Some guys ended up wearing dungarees with white-rimmed holes in them. We learn fast by our mistakes!

Those boys like me, who are very selective about what they eat, learn very soon when it is done the Navy way. My first trip through the chow line for our big meal of the day ended in disaster. I was very choosing of the type of food I would eat, so a big stainless steel tray with six big pockets in it for different kinds of food was picked up, and I proceeded down the chow line.

(Chow Time—Get on the Ball Mac)

The first couple of foods, such as Jell-O salad, and bread was O.K. and so were the mashed potatoes, and meat, but beets were not to my liking so I slid past it: "Get back here sailor!." The mess cook slapped two big helpings of beets on my tray. When I saw the next vegetable was spinach, I thought it was no use to get two vegetables, so I bypassed it, with the same retort from that cook. He too slapped two helpings on my tray.

By this time I was getting the word, that if you don't like something it's better getting only one helping of it by taking it the first offering. My tray was loaded to overflowing by the end of the line.

Then I saw the warning sign and a guard watching. The sign read: "Eat everything on your tray, except bones, or be put on report! signed: Master at Arms!" The guard was posted at the door by the garbage can watching that you obeyed the warning. Wow! I learned fast.

(Learning to Salute)

Our first day on the drill field was a blast! Our drill Instructor lined us all up, showing us how to properly space between each man, and then asked us to salute. I put up a good 3-finger boy scout salute, along with several others, then there were those showing a back-handed British salute, and yet others looking like Indians shading their eyes to look into the distance. What a laugh, and that the instructor did, very heartily, at our expense! Then he demonstrated how it was to be done, and warned of his wrath if we ever did it any other way in the future. We practiced it until our arms felt like dropping off, to the tune of the instructor's laughing; finally snapping a very curtly salute.

He then proceeded to show us how to march in formation, but occasionally added a command such as: "To the rear, march!" We ended up with men jostling each other as we went every direction except to the rear. "To the right flank; to the left flank; forward march; to the rear march; halt one, two!" "Right face; left face; about face!" "Forward march; column right; column left; column to the rear!" These and many other commands he taught us, and practiced them until they became almost automatic to perform.

Some men tried to think what to do and messed up the whole column, by going in the wrong direction. This of course brought down the wrath of the instructor on all of us, a squad of: dummies, idiots, to use his terminology.

In a few days of this though we began to move more like one body, than separatists. Our instructor would even occasionally offer a note of praise, but usually just before another insult when one of us would goof.

This couldn't go on forever, because they had a schedule of five weeks to whip us into shape. So we were introduced to the *obstacle* course!

This was fun for me, because it reminded me of my younger days growing up in the country in Pennsylvania. There we had played Tarzan in the woods, waded creeks, jumped from tree to tree, swung on huge swings, slid down cables strung from tree to tree, swung on ropes out over chasms or ponds, as well as things too numerous to mention.

(The Obstacle Course)

Now the military was to show us how to really do it, and in the process develop our physical stamina in preparation for the rigors of battle. The Obstacle Course was guarded by officers and instructors at each obstacle to see that we did it properly, without killing ourselves.

The sequence went like this:

A Marine stood at the start point with a stop watch, swatting our backside to urge us on, where we met with a log we could either climb over or go under. Then we came to a chasm to cross, with a 12-inch log over it as our path. Next was a muddy slough with overhead dangling ropes which we must jump out to and grab to swing far enough to land on the other side. To fall short coated us with embarrassing mud. Then was a series of short stumps to step on to miss the simulated land mines, followed by a high, solid wooden wall to jump and climb over (this was really a trial for us short guys). We ran then to a tunnel to crawl through, and under a barbed wire barrier at the exit. On to a sloped landing net, 15 feet high, to climb up and then down the backside upside down or else jump off. Another chasm was this time equipped with ropes to swing across, and on to a maze of barbed wire to crawl under. Then to a tree to climb, and cross hand over hand on a rope stretched over rocks below to another tree. Exhausted and out of breath and pouring sweat we exited the course, glad to have survived.

The first time through was merely to acquaint us, but later we would be timed over the course. This was only the beginners course, and in advanced combat training we would have a longer and tougher course that taxed every muscle and nerve in your body, all in a required amount of time.

We found that every four days we had our turn at "Duty Day." When I was given K.P. duty it was delivering meals to the officers as mess clerk. As an officer came through the dining hall door I would take his ticket, and watching where he would choose to sit,

I would get into the chow line at the serving table. I would pick up the meal, beverage, and dessert that he ordered, and return it to the officer. Several of us were assigned this duty and we would take turns greeting the officers at the door. We would each try to serve more meals than the others as competition. This was fun.

At other times I served guard duty, and always had the same officer as Master-at-Arms. He selected me to be roving guard, who drove him in his jeep all throughout Camp Peary, where he would challenge each of the entrance guards, to see if they were properly performing their security duties. He would also surprise some of the interior guards at their posts, and question them on their "Navy Rules of Order." He made it his duty to instruct me on how best to handle myself under Navy discipline, and indicated how much freedom we enlisted men could have within the military limitations. We became good friends, and he would arrange for me to be his jeep driver each time I was on guard duty.

(Our 7th Shot)

During our entire time in training we received periodic shots for all sorts of possible diseases we might encounter. One time just after receiving one of the stronger shots, we had a dress parade on the drill field. The day was terrifically hot and humid, and we were standing at attention too long. Just before being dismissed, I felt as though I was going to pass out, so I knelt on one knee and put my head down, and one of our officers saw what was happening and came to my rescue. While all the other men left he stayed with me, and when I was able to get up, he put his arm around me and assisted me to our barracks. There he had me lie down on the bunk until the faintness disappeared. I thought that was really a kind gesture, and was indicative of the camaraderie we had between officers and enlisted men in our outfit.

As we got better at marching in drill fashion we were equipped with dummy rifles in order to learn how to drill with our rifles, and how to do infiltrating field maneuvers. These looked like actual rifles, and gave us the feel of real firearms, but they wouldn't shoot bullets. In other words we couldn't hurt each other as we were learning the basics.

We were told a gas drill was coming soon, so they issued us our gas masks. This became part of our permanent equipment. Then one day they took us to a Quonset hut and instructed us we were to put on the mask and enter an air lock. Once inside, we were opened into the long building filled with tear gas. We felt immune to gas attack with our masks on, as we pushed each other in the crowd to move through the building. This is a cinch, we thought, and wondered why the others had thought it so traumatic.

Then they closed the exit lock and told us to remove the gas mask. Now we felt the impact of the gas, and it was like pepper had been sprayed in our eyes, with terrible burning, and watering, and noses running, and itching. We pushed and shoved to get to the door which they now opened. Coughing and sputtering we emerged from the building, as true believers of how gas can incapacitate a person, and yet this was only simple tear gas! Anyone who had suffered from broken ear-drums was not allowed to go through this drill.

Finally came the day we were to be commissioned as a genuine Seabee Unit. We dressed in our finest white dress uniforms and proceeded to the Parade ground, amid bands playing, and a large crowd of civilians, families and friends. We felt quite important and jubilant!

Each squad formed up in the best military manner, with dummy rifles, and stood at parade attention. Then the drums rolled, the Marine Band started playing a March, and our marching orders were given. And as proudly as any Unit before us we smartly marched to the martial music, with our hearts feeling as high as the sky! Oh! We felt like no other military outfit could possibly have matched us in smartness and finesse.

What was just a few short weeks before, civilian young men from all parts of America's great land, was now a well trained military unit! Of course, we knew much training still lay ahead, but now we would be a cohesive unit of the newest and best branch of Uncle Sam's Military.

On August 7, 1943 we became official members of the United States Navy Seabees.

(Commissioning Ceremony—110th Navy Construction Battalion)

In front of the stands filled with admiring civilians we came to Parade Rest. The Admiral and other Military dignitaries formed in front of our Officers, and the formal ceremonies began. Finally, we were commissioned the 110th U.S. Navy Construction Battalion. One battalion, with five companies, each with 4 platoons, and several squads in each platoon. The band struck up a Victory March, and we paraded proudly before the cheering and clapping crowd. This was our last formal parade before entering the specific training we needed for the assignment our new Battalion would perform until the war was brought to a victorious conclusion. We had completed five weeks of Boot Camp, and I was now 21 years of age.

Camp Peary—Advanced Training

For our advanced training, we were moved to new barracks, but in the same camp. This area was equipped with more, and different facilities. There were large Quonset huts housing various schools for the different trades represented in our battalion, each with their particular machines.

The first order of business for this new organization, the 110th Construction Battalion, was to find out just what each man's job was to be in this new unit. Eighty percent of the members of this battalion had *enlisted* into the Seabees, rather than be *drafted*, and some of them were Officers, Warrant Officers, others Chiefs, and still others were one of the three Petty Officer ratings. The last two hundred of us were coming in as draftees, with the rating of Apprentice Seamen (lowest on the scale). Yet we all must go through the interview procedure to see where in this big outfit of approximately 1100 men, we could best be utilized. The idea was to fit the person in the job he was best qualified by experience.

Most of the Seamen were men with little or no experience, and the majority of them right out of high school, so I was quite an exception by having three years of industrial experience as a machinist. Because of the particular civilian job I had in the Scientific Laboratory, and rules governing deferments, I was not permitted to enlist, and automatically when released from deferment became classified 1-A. Under the government's rules that meant I could not enlist, but must wait to be drafted. Here I was as a Seaman, who was supposedly unqualified for anything except menial tasks. What could I hope for in assignments? All I could do was pray to the Lord, who knew my situation, and had placed me here. "Wherever you want me, Lord, you open the door!"

The appointment time came for me to meet with the specified interviewing officer, and I entered his office. With my personnel file in front of him, he started asking questions. "Would you like a job in the office? Can you type?" I answered, no, not in the least bit interested. "Well, can you shoot?" My answer was yes, I've owned a rifle since a teenager, and successfully hunted animals. Then he pointed to my machinist experience resume, and indicated it looked quite impressive, but wondered why I was drafted rather than enlisting with a rating. He was then satisfied when I explained the circumstances. He asked me to name the machines I operated and some of the technical tasks performed, and how I liked that kind of work.

Now I was getting really interested in this interview, and it seemed he could sense my enthusiasm. I wondered why he pursued this in such detail, because I knew there were dozens of experienced machinists from the Detroit automobile manufacturing area in our battalion, who had enlisted, all with higher ratings. I didn't dare to get my hopes up too high, only to have them dashed. Before we were through, he told me he was assigning me to the Navy Machinists School and I would be attending the ten day course here on the base, to acquaint me with Navy type machines. He indicated, that according to my file this was the purpose for which I had been drafted into this particular branch, Seabees. At the successful completion of the course he would see to it that I would be attached to the Battalion Machine Shop crew! Wow! What an answer to prayer. After the interview it was like walking on a cloud, as I returned to the barracks and told the other guys.

Several hours each day were spent at the Machinists school, and the rest of the time we trained in other activities. I was exempt from the swimming class, because I passed the good swimmer test, by jumping off the 12 foot tower, and swimming across the pool. We advanced to the more difficult obstacle course and were timed to complete it in 15 minutes, which caused me no problem. We still continued some drilling, but the marching was transferred to hikes across the back lands of Virginia, through the woods, and fields. This was when they started training us to do without much drinking water, even during this hottest time of year. We drank before our hike began at 0800, again at noon, 1200, then at 1500, and finally when arriving back at camp at 1700. (24 hour Navy time) We began to think we must be going to the tropics in the Pacific Theater of war.

We realized we were in a shooting outfit when they issued us our M-1 Springfield rifles which we would use in this phase of training. Then came time on the live firing range. We learned how to clean a rifle, take it apart, and reassemble it, and lastly how to shoot it, hoping we could hit the target. We all eventually got fairly good at hitting the bulls-eye, some better than others, and fortunately I was able to qualify as a "sharpshooter." Evidently my years of target practicing with my rifle as a teenager had helped my aim.

They taught us battlefield tactics of infiltrating, and trench warfare. They even sent us out to test these tactics in a mock battle, in which we infiltrated into a forest and surprised the troops in a trench, capturing all. Our squad even received special commendation from our officers. Our advance scout was a Sioux Indian from North Dakota, by the name of Stead, therefore we had the advantage! One day they took us in landing barges on the James River, and we practiced beach-head landings, wading onto the beach from chest deep water. Alongside me was one of our men who could not swim, and he stepped into a sink-hole and blubbered in over his head, and I had to pull him out. We captured the beach in professional fashion, holding our rifles over our heads to keep them out of the water as we waded ashore.

Another type of training was with machetes, taught by a Filipino. We used wooden ones, but demonstrations were done with sharp steel ones. These are used in jungle warfare. They taught six strokes used in hand to hand combat, which supposedly guarantee total disablement of an enemy, and we got to be quite adept at them as we practiced with our wooden machetes. Then they taught us many moves using Judo to disable an opponent, and then how to kill a man with one stroke of the hand using a Judo technique taught by an oriental man. We certainly didn't try this deadly strike on anyone. Information concerning this instruction was restricted, and took place in an outlying area of the camp.

One of the advantages afforded those in Advanced Training was that we were given "liberty" into town. Before that we were confined to the base. Our liberty town was Williamsburg, Virginia. The first stop in town was the drug store on Main Street, to buy limeades, which were so terrifically thirst quenching. The men would line up along the long counter six or seven deep, waiting to order them. We explored this historic town from end to end. (In later years I was to bring my family here and relive those youthful days).

The close-order drill marching at this stage of training was instructed by Marines. Even to understand their orders was a trial. But Marines were our buddies, and we were glad to get this sharpening up. The first group of Frogmen, Under-water Demolition Teams, (UDT) were being trained on this same field so we watched these men, always doing their marching in double-time (running) dressed only in shorts and shoes. They were all volunteers from the Seabees and as this class developed we could see the fat wear off some, and the skinny ones become husky, and by the end of their training they were all of one size. We heard this outfit was assigned to go into the European beaches on D-Day, to clear away obstacles, and to destroy communications. They were told the casualties were expected to be 60%, and yet they volunteered for the job. They were worthy of admiration! The obstacle course that timed us at fifteen minutes was theirs to do in five minutes. When they went to chow at lunch time they ran to the Olympic sized swimming pool, dove in and swam its length, jumped out and ran to the chow hall, where they had fifteen minutes to eat. Everything was done in fast time! We held them in highest esteem!

At the conclusion of Advanced Training all the men from the Eastern part of the U.S. received their 10 day Pre-Embarkation leave and a week later the rest of us were given a 72 hour pass. They limited the distance allowed to travel in hopes to keep Californians from trying to get home. Remember all travel in those days was by car, bus, or train. Airplanes were for flyboys only. So, traveling with some buddies by train going to New York City, I decided to visit my old hometown of Centermoreland, PA. This was stretching the distance somewhat but it was worth chancing it. The train went to Wilkes Barre, and I caught a bus to Dallas, and started walking the last seven miles, but was picked up by a passing car, and let out only one mile from my destination. To say the least, I surprised everyone in town, and had such a good time impressing the girl friends who met me. These were some of my future pen pals. They all were thrilled to have a uniformed service man in this out of the way place. Most of the men were farm boys who were exempt from the draft. But the time went altogether too fast and I had to start back to camp.

The first hitch was that on Sunday there was no train going to Philadelphia where I must make my connection back to Virginia, so I took a bus. It stopped in every town, of which there were many, and I really worried as the hours bled on that I couldn't make it to camp on time. I fortunately caught a train just in time to Baltimore, then a bus from there. Check-in time was 0800 in the morning, and traveling all night long, I was very sleepy as I rode the packed bus, and standing in the aisle, jammed between other standing passengers I fell asleep standing up! That's hard to imagine, but I was exhausted, having slept very little in the last 72 hours.

I pulled into Camp Peary just at 0800, and found they were giving a couple hours leeway for those who were late. We had no barracks to go to because we had loaded all our belongings into a warehouse, so I climbed up on some of the duffel bags and went to sleep.

I was awakened later by some of the other returnees getting ready to load on the train heading South. We had completed 5 weeks of Advanced Training, and it had been 10 weeks already since we arrived at Camp Peary, and now we were moving out for further training. Just as happened every moving day previously, it was pouring down rain this day also.

We boarded the train heading for Gulfport, Mississippi. Something had changed the plans. We heard that now instead of being an Airfield Construction Battalion, we had been transferred to a Combat Unit with the Marines. This was quite disconcerting, and we hoped our commander could do something to get this corrected, but scuttlebutt had it that he had somehow gotten into disfavor with the higher echelons. All we could do was follow. So we headed for the Combat Range in the swamps of Mississippi!

Gulfport—Combat Training

On our way by train they decided to give us a break when we got to Macon, Georgia. We left the train, and in our now smart style we marched in columns through the Main Street, to the cheers, and waves of crowds along the way and in the windows of the buildings overlooking the broad way. I guess they were impressed with all these young men going to serve their country. They even cheered with a southern accent! I couldn't help but think of all this excitement for a bunch of men who hadn't even been to war yet. Would they cheer as heartily when they saw our young men coming back all scarred and maimed from battles? I hoped they would be as thankful. (And today I know that they were, as our nation had been saved from a vicious enemy!)

As the train slowly made its way along the coast of the Gulf of Mexico, we saw many mansions lining what they call the Gold Coast, belonging to some of the richest families in America. Finally our train pulled into Gulfport, Mississippi, and we caught trucks out to our new home at Camp Hollyday, our barracks being Quonset huts, with 21 men in each. This was quite a different type camp from the one we had left, with modern two-story, white barracks which we still didn't get to inhabit. We must first serve our time in the Quonset huts. This camp had a theater, chapel, and P.X. store. We would be here 5 1/2 weeks, and they started us out with a little drilling for discipline, but they had several surprises in store for us.

My first liberty was spent in the homey little town of Gulfport. Many of our other men went to the town of Biloxi, where they started so many drunken fights that the Military Police declared it off-limits to our battalion in the future. But my liberty was spent getting acquainted with the southern hospitality. Several of us went to the outdoor roller-skating rink where we met some of the young southern belles, and enjoyed skating. Afterward I went to a home-style restaurant and sat at the counter. I asked the waitress if the Southern Fried Chicken was as delicious as us Northerners had heard, so she brought me a platter heaped with the most delicious chicken I ever tasted. Ummm! On another liberty I went swimming in the Gulf by diving off one of the large piers, knifing through a school of fish on the way in, and basking in the warm waters, enjoying the sunny flavor of the South.

Back at camp we had been issued all the combat equipment we would be taking overseas with us, including our new 30-Caliber Carbines. They could shoot 15 bullets from the clip, as fast as you could pull the trigger. Of course we had to practice on the rifle range, and re-qualify with these lightweight versatile weapons, and also shoot with the 45-caliber revolver. One trick we learned with our Carbines, when we had it slung from our shoulder, we could swing it around and fire quickly from the hip in gunfighter fashion.

Then one night after a grueling day the lights were turned out at 2200 (10 P.M.), and we settled down for a regular night's sleep. Unexpectedly, at two in the morning the bugle sounded, whistles blew to get us out of our cots. We were heading out, with full pack, for a hike! Our destination was the live-fire Combat Range 35 miles north, in the De Soto National Forest area!

The night was very chilly, but clear, as we headed out into the country. Man, those packs seemed heavy, but only weighted something less than 75 lbs. We were to be gone for some time and these belongings we each carried were our living gear. This was the final training before we actually lived it in the combat zones, for real. Pitch dark, with no moon in sight, we marched on, often to the accompaniment of songs sung in cadence such as: "Red Sails in the Sunset," "We're the Seabees of the Navy," "Swingin thru the Riggin," etc. But that would get tiring, and we just plodded along in the dark, along the country road. After an hour they passed the word we had hiked five miles and would take a 15 minute break. We laid down on the gravel alongside the road, and immediately went to sleep. The whistle sounded and we stiffly wiggled to our feet, struggling to get the backpack into position, and managed to get our legs moving again down the road. For the first hundred yards or so it was a struggle to get the legs to go, but then the stiffness wore off and we were O.K. again.

(Bivouac Area)

This procedure of marching 5 miles, and resting 15 minutes continued on through the night. At about 5:30 am we arrived at what they called the Bivouac Area, where we would rest and have breakfast, and we could soothe our aching feet in the cool stream flowing through the encampment. I took my boots and socks off and soaked my feet in the cold water, and oh, what a relief. I had a couple blisters that had broken open, but my feet were nothing compared to some of the fellows who were almost crippled with blistered feet.

They had announced to us that a truck would pick up any stragglers, or any that felt they couldn't continue on, but that those men would have K.P. duty the entire time we were on the Range. Some of the men were ready to drop out, no matter what the punishment was, but not me. I determined that this Christian could take anything they could throw at us, and I would see the hike through to the finish! If a Christian couldn't take it, with Christ on his side, who could?

They had tents with bare cots in them, so I sought one and laid down only to drop off to sleep immediately. Some time later after the hot breakfast was ready, they blew the whistle, and I awoke, feeling much better. After a good solid breakfast, we all felt much recovered, and realized we only had half as far to walk as when we started. Trucks pulled out with all the stragglers, but they realized there would be many more dropping out along the way now, so a truck was always waiting for any further dropouts. The morning dragged on as we put miles beneath our feet, and finally we struggled into the Combat Camp, our destination in the forest.

It was a great relief to find that all the stragglers that had ridden in the trucks had our dinner ready when we arrived, and oh boy, the food tasted good! Now to set up camp, and get a good night's rest. We really did sleep well that night, like exhausted troops!

The scene was the *combat* range, so everything was planned to simulate battlefield conditions, with live ammunition and enemy fire! We were told the battle plan today was to advance across the open valley and infiltrate into the woods, where the advance scouting party, who had gone out at dawn, had discovered the enemy was hiding. The enemy position was above us, and if we were detected they could shell us while we were in the open. In the woods it would be a total surprise, as the enemy would probably try to ambush us. When we saw the enemy we were to shoot to kill. We didn't think our officers *really* expected us to kill the "enemy," but these were our orders, and we were taught to obey.

We lined up by platoons and squads, with our squad having two Thompson Machine guns, two Browning Automatic Rifles, and the rest of us were riflemen. I was also the ammunition runner, trained to carry heavy canisters of rifle ammunition, while running at top speed in a crouched position, avoiding enemy fire. From our hidden position in a wooded area, we advanced on our bellies through the tall grass, advancing occasionally in a crouched position a few steps, then flopping on our belly's again, all upon hand motions from our squad leader.

The advance line would be kept ragged but moving in a straight line across the valley. All was quiet at first, and we advanced to the lowest part of the valley, evidently not spotted by the enemy.

Then all at once it seemed like the field erupted, with explosions in various places in the field. I thought they must be land mines, but then I heard a whistling sound and one landed near our squad. Our advance suddenly became faster as we went up the hill hoping to make it to shelter of the woods. In the excitement I flopped too hard as I "hit the deck," and my mouth collided with my rifle, and I almost lost some teeth, blood oozing from my cut lip. Then an explosion went off only about ten feet to the side of me, and I thought they are playing too rough! We were making good time up the hill, with explosions going off all around us, to keep us moving.

Our squad made it without any casualties into the edge of the woods. I had lost sight of our Indian scout, Stead, as he infiltrated ahead of us. Then we saw some of our officers directing the individual squads into their positions, and were told to keep on our belly's and to keep our butts low. We soon learned why, as live machine-gun bullets whizzed just above our bodies from the side. We advanced a few yards to some felled trees, where we were to take our stand. We suddenly saw Japanese body-shaped targets pop up about 50 yards in front of us and we prepared to fire, when up pops our Indian scout, right amongst the targets, wildly waving his arms. Our officer shouted to hold our fire, as Stead came running back to our firing line. He had been so sly in his maneuvering, he had infiltrated right past the officers without being seen! We then opened fire on the enemy targets, and knocked every one of them down as they appeared. That was quite an exciting simulation battle! We now knew what it felt like to have real ammunition whistling around our heads, and explosives going off at our side.

Other simulation activities occupied several days, and we began to feel like invincible warriors. In fact the tricks, and games we played on each other were aimed at trying to really hurt the other guy, but we were all too tough to feel it, or to admit that it hurt! This was really a hardening of fighting men, all being urged on by our Marine combat instructors, who themselves already were veterans of some Pacific campaigns, and were back in the States for rest. We *secretly* admired them, but they allowed no sentiment to enter their training. Emotions, yes! But only the kind that would make you grit your teeth and swear to take it out on the first Jap soldier you encountered! After all, he's the one who started it!

They showed us some actual combat films, viewing the blood and guts of real war. They explained our emergency medical kit to us, together with the films. They explained how to stuff the guts back into a fellow fighting man who had been hit by a shell in the stomach. Then to load the wound with sulfa drug powder, and tape the wound together. Then to give him a shot of Morphine, and mark his forehead with a big M, indicating he had already had his morphine shot. This type of training showed us we were in this war for keeps. It was no patsy game!

While at this range, we even had a couple hours each day to enjoy this swampland and forest of Mississippi with recreation. A bunch of us went swimming in a nice stream, with weeping willow trees hanging over the water. We were enjoying this refreshingly cool swim when suddenly someone called out: "A snake!" And as they pointed to us, we saw a six-foot water Moccasin swimming right for us from the mossy bank. I'll guarantee his body was at least 3-inches in diameter. Faster than we could swim we made it out of the water, to escape that deadly poisonous monster. That ended our water frolics for the evening.

Speaking of snakes, I encountered a surprise during one of our infiltrating maneuvers. As we were advancing one man at a time through a supposed mine field, I crouched low as I ran the ten yards in swampy, oozy soil, and flopped on my belly, satisfied I had made it safely. I detected a movement by my chest, and looking down I saw the wriggling tail of a snake sticking out from under me. Quicker than my squad leader ever saw one of his men jump, I leaped up and ran a hopping ten yards in another direction! I never knew what kind of snake that was, but he was probably as glad to get rid of that heavy weight on him as I was to get out of his sight! Who has more fun than me? This felt like just another hardening for the rigors of war.

Boy, at this point in our training I felt nothing could harm me, I was invincible, especially with the Lord on my side! This was probably the high point of my physical stamina. And that is the way our combat instructors wanted us to feel, prior to what they knew would be some very hard times as we met the enemy in the Pacific.

The day came for us to hike the 35 miles back to civilization again, and by this time it was not such an ordeal. As we arrived back at camp we were ordered to pick up our cots and living gear, and move into the 2-story new barracks. Here we experienced what real barracks life is all about with a Marine to wake us in the mornings with a bugle and whistle. We had to be out for calisthenics in 5-minutes.

I've seen a man lie in his sack for two minutes and get a night stick whacked over his feet by the Sergeant. That hurts! He usually jumps out in one leap, even from the top bunk.

At four o'clock one morning, we were awakened, and told we were heading for an invasion of an island in the Gulf of Mexico! "Load your backpack for a stay." It was quite chilly, and foggy as we hiked with our 75 lb. packs the five miles to the dock where we loaded onto a Seabee pontoon barge, and headed out into the open sea. The waves were quite high and splashed over the flat barge, splashing us with cold salt water. We huddled together for warmth, chilled to the bone! After a few hours, with the sun now warming our wet clothes, we came into sight of our Island X, Cat Island!

We were supposed to surprise the "enemy," but instead our pontoon barge grounded on a sandbar a ways off shore. We had to wait until the "enemy" brought a bulldozer and hooked onto our barge and pulled us off the sandbar. I don't think we would have been treated so kindly by a real enemy in battle! Our commander was quite embarrassed, and took it out on the barge pilot.

(Arriving at Cat Island)

We finally landed and proceeded to our encampment spot, since the "invasion" was no longer a surprise. Once our pup tents were set up, we were given time to go swimming in the Gulf. With only our swim trunks on, and no sunscreen (it had not been invented yet), we frolicked on the white sand with games, alternated with dips in the water.

This went on for several hours, interspersed only with a simple meal. By afternoon we all began to look like red lobsters, with sunburn on most every part of our bodies. My belly was red as was the front and back of my knees, and my back and shoulders were starting to blister. We weren't allowed to go to the dispensary, because getting sunburned is against Navy regulations, and can end in disciplinary action. So all we could do was cover it up. That night while trying to lie on the sand in my tent, the pain was excruciating, no matter in what position I tried to lay.

At midnight we were quietly awakened, and told the enemy was trying to sneak up on us, and we had to intercept them somewhere out there in the dark, with our combat pack on our back. Can you imagine how it felt to crawl on sunburned knees, slithering on a blistered belly, over the sand, with a loosely fitting pack on our blistered back rubbing back and forth. It was as painful as any combat situation we would encounter in our war in the Pacific. Finally when the alert ended and no enemy was encountered, we were ordered to separate into individual patrols guarding the beach all night, on alert to the possible enemy attack from the sea. We had no idea what kind of simulation battles they had planned for us, as they tried to make it very real conditions.

I was assigned a quarter-mile stretch of smooth beach, and by this time the moon had come up and the water and beach were bright with moonlight. I patrolled back and forth, and then noticed a dolphin jumping just offshore from me, and he stopped with his head held above the water, evidently watching me. Being fascinated, I talked to him and he responded by swimming quite close to the beach, following me as I walked.

(My Playful Pet Dolphin)

I found a large driftwood stick on the sand and thought I would play ball with him, by throwing it out in the water near him. To my amazement he swam to it and with a flick of his nose flung it accurately back to me on the beach! Remarkable! I tried it again, and he repeated his feat, by again throwing it to where I was now walking. I encouraged him some more by talking to him, and threw the stick several times as before. He seemed to be enjoying the game, and was like a trained dog the way he would retrieve and throw it back to me.

This game continued along the beach for several hundred yards, and now I knew the dolphin was enjoying it. Then one throw from me went wrong, and hit him right on the nose! He shook his head, seemingly in pain, squeaked, and leaving the stick, headed indignantly right for me on the shore at top speed. I jumped back away from the water as he flopped and wriggled on the sand half out of the water, and then turned and disappeared underwater heading out to sea, never to return to me again.

I felt terrible that I had injured such a beautiful creature of the sea. But there wasn't anything I could do now, except ask my Heavenly Father to forgive me for injuring the lovely creature He had sent to minister to my lonely vigil. I learned a real lesson right then about other creatures; how to appreciate them to the fullest. They are a remarkable part of God's wonderful creation!

I also learned later how the Southern boys cooked crabs. We caught lots of them in the brackish swamp there, and heating a can of water they would drop the live crabs directly into the boiling water and cook them. I thought it was cruel, but they assured me it was the way chefs prepared crabmeat. I tried to eat them, but couldn't quite get over the horror of their cooking method. The irony of it was, that here I was being trained to kill enemy human beings and it was even a shock for me to take the life of a little creature like a crab.

Our invasion training ended and we loaded onto the pontoon barge to head back to the mainland. The scuttlebutt now said we were shortly to head West. The word had come down to us (it could have been pure scuttle-butt) that our Commander had gotten the War Department straightened out about our assignment. As a civilian contractor who had built air fields, he let them know, he had enlisted to build airfields and not to command a combat unit! So the mis-assignment was corrected and we were again destined to be an airfield construction outfit. We no longer would need further combat training but needed to learn a little more of the actual construction work we would be doing.

We must go to Hawaii in order to actually perform construction, because the labor unions would not allow such work to be done in the States, as it would take civilian jobs from the unionized construction trades. So we weren't even allowed to do any of the construction, or trades work on our bases, in the States. Hawaii was not yet a State, but was the Territory of Hawaii. A unionized construction of an airfield in the States usually took at least three years, and we found we could do the same job in a few weeks, or months oversees, with Seabees doing the work! Such is the workings of a democracy, rather than a totalitarian State.

The advantage of a Seabees outfit was that it didn't require nearly as much time to train the men to do their particular military job. The majority of men enlisted for this service, and they were already experienced in doing what they would be required in service. Consequently we were shipped overseas in a very short time, with only the military orientation, and physical stamina training behind us. The actual combat experience was of course still ahead of us!

Back at Camp Hollyday we prepared to move again, this time to the West where we expected to depart for overseas. Of course, we never got very attached to any of our homes in the service, as we knew it was only temporary, until we would move on again.

The battalion was divided into three echelons for train movement, so as to confuse the enemy spy system, and our particular section was to go on the most Southern route of the Union Pacific Railroad. This time we had real Pullman cars, and berths! "Move em out!"—Was the order we again heard as we boarded the train for somewhere in California!

(Young Men Heading West Again)

Port Hueneme—California Here we Come

Those of us from the West Coast were overjoyed to be on our way, because this meant we would have our 10-day Pre-Embarkation leave soon. This train trip also broke the stress of combat training, as it had been quite an ordeal, and took a lot out of us. Now we could rest! The countryside passed by and soon we entered the huge state of Texas. That seemed to never pass by! We couldn't imagine any state could be this big, with it's variety landscape of hills, trees, lakes, small towns, and endless prairie flatland.

Often we passed by passenger trains sitting on a siding to let our troop train through without delay. We had plenty of time to think of what lay in our future, though we thought more of what the next few months held for us rather than long-term. Little did I know that 30 years later all train movements on this very same Southern Pacific rail line would be controlled by a telecommunications system for which I had been Product Managing Engineer for its development, a GTE Lenkurt 36A System. Our thoughts at that time were more concerned with surviving the war; and then we would have time to plan beyond.

At that present stage of the game each of us simply lived each day as it arrived and the military brass did all the planning, and hoped also to do our thinking for us. They supplied all our needs, like parents take care of their children. In fact some of the fellows seemed to thrive on this life of being cared-for, but my independence was never squelched. Personally, I thrived on the adventure, the unknown risk factors, and learning aspects of the life we found ourselves in, and always wanted to expand my vision and interests. I was very curious; the kind that sometimes gets the cat in trouble!

We Californians had painted quite a rosy picture of the beauty our State held in store for us, so the Easterners had great anticipation! Their thought balloons were burst as we crossed the state border into California right through the driest, emptiest, hottest part of the Mojave Desert! Their spirits hit rock bottom. We tried our best to encourage them that soon we would be entering the beautiful part of Southern Cal, but they only insisted we were nothing but a bunch of liars. Only then did I realize much of the enjoyment of our Golden State was in the mind's eye of the beholder, and specific places, people, and activities available. But this was my home! I had planned to toss a note out the window as we passed through San Dimas, my city, but they took another route—disappointed!

Again, many hours were spent in the Los Angeles train switching yards, which certainly is not a place of beauty, but finally heading North to our destination of Hueneme, at about dusk. As we approached the beach area of the Pacific, we entered a chilly fog. We were blanketed with fog the rest of the way, unable to see any of the sights. Arriving at our destination on October 22, 1943, after a four day train trip, we stepped from the warm Pullman car into the damp chill of coastal fog. We had our p-coats (overcoats), and they were indeed welcome. Again our friends, ridiculed us for telling them they were in Sunny Southern California. "Liars!"

(Fond Farewell)

(Doug, Dennis, Uncle Clay)

We were put up in single-story barracks, and told to make ourselves at home with a good night's sleep, which was welcome. As soon as all battalion personnel were organized, they would release us for our 10-day leaves. On October 25, we were sent on our way, and ordered to be back by November 4, 1943.

Our family had several get-togethers during the next few days, with picture taking in uniform. It was at this time the portrait of a Seabee was taken which now decks the Introduction page of this manuscript.

My nephew, Dean had been born recently to my brother, Lester, and Jean, so I visited them, and then Les took me on a tour of the place he worked, Hollywood Christian Servicemen's Center. Lester led all the services, and brother, Alvin played the piano for all the singspirations. I vowed to bring some of my buddies there later. Being the only one of our immediate family in service, the others thought of me as their representative in the military.

My brothers were all too old to be in the draft, and besides they were each in deferred occupations. Lester, and Alvin were ministers, and Marius was an important link on the P-38 fighter plane development and production at Lockheed. Sister, Gladys was working at Lockheed on the manufacturing of airplanes for the military. (Typical "Rosie the Riveter"). Dad was Pastor of the San Dimas Baptist Church, and Mother was his worthy support. Dad was also a "Block Air Raid Warden." So they all gave me a great send-off! The leave was over all too soon, and my diary records that I had a *swell* time of it! So back to camp I went, to see what other assignments were forth-coming.

Upon arrival at camp I was notified my rank had been upped to Seaman 2nd Class, with pay of $56 per month. I was assigned to the Chief's Mess Hall, for the duration of our stay. That duty required us to be on call for 12 hours of each day, but only working the hours around two meals of the day. Our shifts covered breakfast and lunch, or dinner and next breakfast, so that we had 12 hours liberties at least every other day. The Chief's all had to pay for their food, and were accordingly given food allowances, the same as Officers. This meant they wanted the best of food, and consequently those of us who served there received the same kind of meals. All leftovers were divided between us, which meant we always had cake, pies, and fancy desserts. When leaving on liberty, we would make sandwiches, to take along, of complete steaks or chops between two slices of bread, dessert of our choosing, and fresh fruit. So we didn't even need to buy food while off the Base. We felt we were eating like kings!

Twelve hours was not a comfortable length of time to go all the way home, especially with the public transportation available, so most of the time Bill Dally and I would headquarter at his parent's house in El Segundo. From there we went bowling, or swimming in their city indoor pool, where he knew all the pretty girls from his school. I met his sister, Betty, and we enjoyed times together. We would sometimes go to Hollywood to see the sights, or to visit the Hollywood canteen.

Also on the strip was the Hollywood Christian Servicemen's Center where my brothers ministered. Every day there were churches whose women worked at the canteen counter and who provided cakes and baked goodies for the men to eat with coffee, and soft drinks, as well as providing a taste of home and conversation.

(Hollywood Christian Servicemens Center)

I had bragged to several of my buddies about my sister-in-law, Ruth's wonderful chocolate cakes, with special whipped frosting. No one else could make one like she does! The fellows didn't believe anything could be that good. Well, I invited several of them to go with me to the Canteen, hoping that maybe Ruth would be there with one of her cakes. As only the Lord could provide, she was there and had her special cakes in hiding for a special occasion. She served it to each of us, and Oh! How the guys raved about that delicious cake. I think my Christian testimony was solid gold after that, when they knew I spoke the truth. They all were terrifically impressed with the service, and singspiration, that went along with our visit.

Afterward, the fellows could not get over the fact I had brothers with such an important public service, in the city of Hollywood, and a brother that had a wife who was such a terrific baker! Of course, during my entire time in service I was sharing with others what a wonderful family I had, and what they were all doing! I even had the opportunity to share this with the Captain in charge of the entire Atoll of Eniwetok, as you can read in my story of the "Barge Sinking."

We sometimes stretched our liberties a little too thin on time allotment, and sweated it out on our chances of making it back to camp on time. One time I stayed overnight at my Aunt's house in Los Angeles, and my parents were there also. I was due back at camp at eight in the morning, and Dad was to drop me off at a bus station about half way to camp. When we were almost to the bus terminal I discovered my wallet was missing. It must have dropped from my pocket at my Aunt's house, and it had my I.D. pass in it.

Dad whirled the old Model A around, and we raced back. Finding the wallet, we saw there was not enough time to get to camp unless Dad drove me all the way. He had the throttle to the floor all the way up to camp, about sixty miles, and we made it to the gate just in time! I had never seen that car go so fast.

Another time we thought of spending the night at the Venice amusement park, and when it closed at midnight we had no place to stay. We curled up on one of the park benches along with some of the bums, wrapping newspapers around us to keep out the chill, foggy wind, (like we saw the bums do). However after a couple hours, being chilled to the bone, we gave up and started back to camp, where at least we had a warm bunk. We tried hitch-hiking but with very little traffic, and with so many servicemen doing the same thing, we were unsuccessful in getting a ride.

When a Greyhound bus came along we hailed it down, and paid the driver to take us with him as far as Ventura. Pushing through the standing servicemen in the aisle, I spotted a place in the upper luggage rack that was empty. Instructing the guys to wake me when we arrived, I climbed up into the rack and amidst the pillows there went right to sleep, warm as toast! At our stop we got a service bus to camp and into our own bunks almost at daylight.

Being so very tired, I laid on my top bunk with all my clothes on. In the morning, while I was still sound asleep, some of the guys thought of having some fun. So they stuck a match into my dress shoe sole, and lit it. I woke suddenly with a hot-foot! No one would admit doing the trick, but they all looked guilty. It's all in the game!

As they were making up the military unit that would invade Island X, our 110th Battalion was assigned to ACORN 22. This acronym stands for Aircraft, Communications, Ordinance, Radio, and Navy. This unit would provide all the support functions for the invading forces. This was accompanied with a dress parade, and commissioning, but at the time I was on duty, and did not participate in the ceremony.

The men who were late in arriving back at camp after their 10-day leaves were transferred out of our outfit, so I lost one of my friends, who was then sent to Alaska. The scout of our Platoon 3, Stead, came to me one night and poured out his heart about how badly he was being treated by our officers because he was a Sioux Indian. He planned to desert, and return to his tribe in the Dakota's.

I tried to talk him out of it, but he insisted they couldn't do anything about it because he belonged to another nation, and therefore was not a citizen, and couldn't be imprisoned.

The next morning I came into the Headquarters officers building while our Commander was really cussing-out our Company-A officer for driving Stead, one of our best heavy equipment operators, into deserting. They had a note from Stead which described his situation, and he had told them not to look for him, because he was going back to his own Sioux Indian Nation. He was disgusted with the treatment by white men, when all he had wanted to do was help win the war. That officer took quite a lashing! I really felt badly about my friend, Stead.

Arrangements were all completed and we were confined to the base in preparation for our embarkation. Our first stop was to be Hawaii, where we would train and prepare further for the actual work we would be doing after we invaded Island "X." Good-bye's had all been said to our loved ones, and we were ready for the "adventure" of a lifetime! We boarded our troop ship on November 21, 1943, with our seabags, packs, masks, and rifles.

We were in the bottom hold of the ship, with bunks five (5) high, jam packed! We didn't leave until the next day, and at 2:30 P.M. we pulled away from the dock! That was a time to be remembered for life! As we cruised out of the harbor, and headed for the open sea, we watched the land disappearing from sight, and what a tugging on our heart!

(Boarding the Liberty Ship)

(Assigned Top Bunk, in Bottom Hold.)

We were leaving the good old USA and wondering when we would see her again,—if ever! But despite the physical emotions, I was very confident of returning safe and sound by answer to prayer and it being the Lord's will. We all stood on the weather deck to watch the land disappear. As it got dark we could still see the tops of the mountains of the California coast. Oh, man! Oh man!

Before long a good share of the 1500 men aboard ship were hanging over the rail, "feeding the fish." My stomach felt queezy, so I went to my bunk. About eight in the evening, I gave up the job, hustling up on deck and joined the others "hanging over the rail"! Feeling much better, I climbed up to my bunk and went to sleep.

Next morning was a repeat of the night before, but now I didn't get out of bed, but rather, used my steel helmet as a bucket for my seasickness. I stayed in bed all day, eating nothing, and you may imagine the smell in the hold with most of the men seasick. Next day I stayed in bed and ate nothing. The following day it was the same as the day before, not even getting out of bed, but a buddie brought a sandwich to me! On the fourth day I managed to get out and eat breakfast, but stayed in bed all afternoon. On the fifth day I managed to eat breakfast again and stayed on deck all day. I bought some snacks from the ship's store, and ate it. On the sixth day, I ate both meals, but didn't feel very good, and had a sick headache.

Next day I felt fairly good and laid around on deck all day. Saw schools of flying fish. Every other day we had “abandon ship” drills.

By the eighth day I had gotten my “sea-legs” and was really enjoying the rocking and rolling of the ship! The water was very rough the entire trip. We were all anxious to see the sight of land again, and on December 1, we hove into sight of Diamond Head on the island of Oahu. What an inspiring sight! Planes kept flying around us saluting our convoy. One ship at a time was allowed to pass through the submarine nets, as we entered Pearl Harbor. We docked in a branch of Pearl Harbor just opposite Hickam Field, while it was raining, and then we saw our first tropical double rainbow, brilliantly displayed in the sky. It was beautiful!

Other troops disembarked, but they kept our battalion awhile longer, as we were to use the ship as our quarters while they prepared a place for us to move into. We were able to go ashore to shower and to visit the store. I worked on K.P. duty for a couple of days, then they assigned a permanent detail to it. We finally moved off the ship on December 5, and into Quonset huts at Iroquois Point.

On the anniversary of Pearl Harbor Day, Dec. 7, Trigg and I went over to Hickam Field, and did some sightseeing. It had been exactly two years since the infamous attack on Hawaii, and we saw the large hanger that was destroyed during the attack. It was still a shambles. We saw some Hawaiians, and lots of Japanese, who all worked on the military base. We still had our I.D. Cards which are what they use for liberty passes. We were restricted to the base, and we heard they were going to take our I.D.’s away from us so we couldn’t go off the base, and would give them to us only for official liberties.

We hadn’t been assigned to a permanent detail yet, so Trigg and I planned to do some sightseeing in Honolulu, before we lost our I.D.’s. Each day a new batch of names would be posted, designating men for various details. If we left the Base, and our name was posted, we were in trouble.

Our plot was strategized. We would be absent without leave, but we would leave early in the morning, December 8, before the duty list was posted. We would still be legal as far as the gate Marine Guards were concerned, because we had our I.D. cards. We now knew how to get the taxi-boat to Hickam Field. There we went to a gasoline service station, and asking the owner if we could change clothes. We took off our fatigues, changed into our dress whites which we had rolled up in a paper bag, and we walked to the bus stop at the gate. We had left our fatigue clothes in a bag with the station man, trusting him not to play a trick on us.

Near the exit gate from Hickam Field, we boarded the bus, all the time feeling like spies trying to get away. As the bus came to the gate, Marine guards boarded to check out all I.D.'s. What if they had orders that no Seabees were yet allowed to go out on liberty? We were scared! We passed inspection, and the bus pulled out for Honolulu with two suspicious looking Seabees aboard.

Getting off the bus in the city, we visited all the souvenir shops, picking up post cards and trinkets. I also was brazen enough to have my picture taken with a native girl in a hula skirt.. The native girls were very attractive! The place was swarming with servicemen, and especially sailors. The cities were on curfew after 6 P.M., so all liberties ended at that time.

We took a bus to Waikiki, and while we were trying to open a coconut that we had found, a native man drove up to us and asked if we would like to be shown how the natives do it. After giving us a lesson, and demonstration he offered to take us site-seeing in his car. He took us on a tour of the entire city, explaining what all the places were, and the buildings. It turned out he was a Christian, and Pastor of a local native church. He gave us a rundown on the Hawaiian culture, and what they expect in courtesy from the Americans, and servicemen. We had a marvelous time with him, and as he let us out by the Royal Hawaiian Hotel, he got out and as he shook hands left a dollar bill in the hand of each of us. He thanked us for sharing the blessing of Christian fellowship with a native. This was a highlight of our time in Hawaii!

After touring through the beautiful hotel, we caught the bus back to Honolulu and then to Hickam Field. All went well in getting back to our base, and we were relieved to find our names had not been posted on the duty roster. However the next day we were each assigned duty, and our I.D. cards were taken away. What a narrow escape we had! This outing to the city had to suffice for another week before we got a legal Liberty.

While on my next Liberty I bought a cheap camera, to get some record of what I was seeing, and to get pictures of my buddies. I recorded some native activities, such as kids doing coin diving in the canal at Waikiki, and native women weaving flower laies. I was invited to join a group of native college girls on the reef doing biological studies of the underwater life, using glass bottomed boxes to view Marine life on the coral off Waikiki beach. At this time, the only hotels along the beach were the Royal Hawaiian, and the Moana, in which many of the wealthy families of the thirties, stayed. We enjoyed sampling the view in their lobbies, which were elegant!

(Our Christmas Hula Show in Hawaii)

Back at the Base, we had a very unique Christmas Eve, being entertained by a U.S.O. Native Dance team. We even got to see "Hula Hattie" do her famous "Hula Hop" of which the song is written: "When Hula Hattie does her Hula Hop!" It was really something to write home about, but of course we could not describe it because all our mail was then censored, to cut out any military information that might give enemy intelligence knowledge of troop movements or activities. All our folks knew was that we were somewhere in the Pacific.

On January 7, 1944 I received my official transfer from general duties, to the Battalion Machine Shop. This was the culmination of the assignment made when I was interviewed for Advance Training; it had now finally become a reality. I reported there for duty! On February 1, I received an advancement to Seaman-First Class (S-1/C), having passed my orientation as a Machinist.

Further revelations are to be revealed in the following stories, which were not permitted to be told until after the war when we had returned home. Thus our fervent training was now about to be transformed into ACTION!

> "All of life is made of a series of steps, as we grow by our experiences. Our plans often seem to take sidesteps, but we are still going in the same direction. Each of us is different, yet together we make up the whole. As we work together toward a common goal, amazing things can be accomplished, but we need a leader who knows the way! Like a flock of sheep we may get lost, and we need a shepherd to guide us."
>
> The Lord is my Shepherd!
>
> —A VET

(Our Vision if We had Only Known)

Following our intensive five months of training in the States we were now prepared to face the real battle in the overseas conflict. Travel with me as we experience some of the events that helped shape my life. Remember, we did not know from one day to the next what lay ahead, we just lived it!

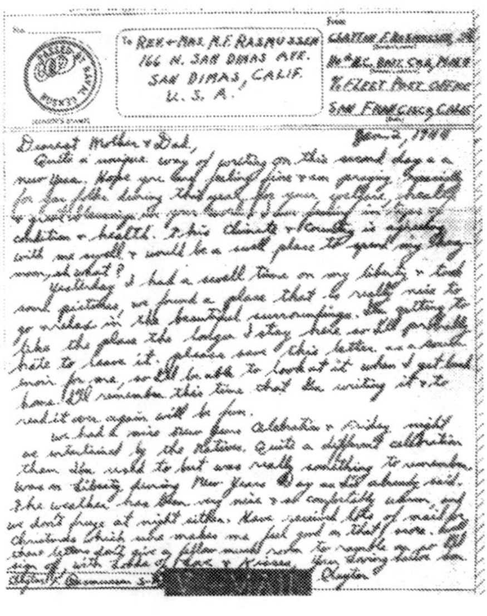

To REV. & MRS. H. F. RASMUSSEN
166 N. SAN DIMAS AVE.
SAN DIMAS, CALIF.
U. S. A.

SAN FRANCISCO, CALIF.

Dearest Mother & Dad,

[illegible]

Clayton

[redacted]

(Our New Form of V-Mail While Overseas)

ISLAND "X"
INVASION
"MOP-UP"
SQUAD
ENIWETOK

Chapter 2

Island "X"—Invasion "Mop-up" Squad

War is horrible! And the experience can be devastating to a young man. Yet faith and trust in our Savior, Jesus Christ, can provide safety while destruction churns around you; and peace in Him, even while shells are bursting, and all manner of danger lurks at every turn. These experiences help develop our faith.

—A VET

What a vivid imagination youth has. Many were the teen-age daydreams I had of traveling to the beautiful South Sea Islands; swimming in the azure blue lagoons; falling under the enchanting spell of dancing Hula girls; surfing in the warm, gentle waves with tropical fish. And, now, here we were by courtesy of good old Uncle Sam, on an all-expense paid tour of those very islands, read so much about in writings by Robert Louis Stevenson, Howard Pease, and Nordhoff and Hall. True, we had been thrilled with our first experiences here in the Hawaiian Islands, and it was as enchanting as the daydreams, but a shadow hung over this present picture. Our nation was at war; against a cruel and merciless enemy. We were here temporarily to finish preparations for an onslaught on one of the enemy strongholds, in the Marshall Islands.

The air of expectancy, and "turn me loose, on those Nips" attitude, was growing as we entered our third month in Hawaii. Even so much so, that my friend, Bill Dally and I volunteered to join a hand picked squad, made up of Navy swimmers, that was being formed by the Special Forces. This was for a landing on an undisclosed beach in the Pacific, ahead of the main invasion force. It probably was a special Frogman mission. If picked, we would train on another of the islands of Hawaii, using new tactics especially adapted to this mission. But that operation was not to include us, because our battalion moved out before the unit was formed.

I had volunteered to serve one entire month on K. P. duty in the kitchens, so as not to have to serve it again in the future. I could then serve all my time in the Machine Shop. One advantage on K. P. duty was to get the best of food and desserts. Little did we know this was by far the best food we would eat for the next two years. Fortunately for me, I didn't enjoy eating fish, so I didn't eat any the day one-third of the battalion got food poisoning from the fish served. We had ambulances coming from all directions, picking up guys as they lay in agony on the ground, and in their bunks. What excitement!

Then, one day when we least expected it, all liberties were canceled. A freeze was put on all mail going out or coming in. The word came our Battalion was moving out to Island "X." Rifles all cleaned, cameras confiscated, our equipment all checked, bayonets sharpened and oiled, and postal cards to fill out as to our condition after the invasion. One card read: "We saw action, but I am well." Another said: "Am in the hospital, but am recuperating well." The last one was our will and testament, leaving our personal effects to 'so and so.'" (That one was for use if we were killed in the line of duty.) These were addressed, and put on file in the

Headquarters office. Boy, you can imagine how this made us feel, not knowing which card would be sent home!

A few of us Christians met, and with H. Trigg as our leader we decided to distribute Christian tracts to every enlisted man of our battalion. We chose the tract, "Do You Know Where YOU Will Spend Eternity?" The Navigator's Club gave us enough to go around. Most everyone accepted them and read them, with deep concern. All were thinking about the possibility of not coming back alive! This was a difficult type of Christian witness for me to do, but when I saw the sincere thankfulness on the faces of some of the men, it strengthened my faith. We were able to uncover more true Christians, who later joined us for prayer and Bible Study.

I was called to the Shop to help seal up all tools, and to put preservative on them to prevent corrosion from humidity and salt air in shipping. We locked down all equipment in the Shop Trailer, ready to open and put into immediate service at our destination. It was at this time I was notified that my rate had been advanced to "Seaman, 1st Class," with an increase of pay.

Our Commander had found out the Navy Supply Depot was sending us on an invasion, to build an airstrip, with only 50% of our required heavy equipment. **He was furious**! In civilian life he was the owner and contractor of a large company whose job had been to build airports. He knew what it required! A job, that in civilian life would take three years, he was being ordered to do in a matter of a few weeks. So, he put into operation what the Seabees call, "Midnight Requisitioning." He signed a bunch of requisition orders for bulldozers, cranes, scrapers, loaders, trenchers, trucks, Jeeps, and many smaller items for support. He gave each requisition to a different heavy equipment operator, and sent them to the depot all throughout the night. The operator would drive his equipment to the dock, and they would be immediately loaded aboard the ship, into the hold, out of sight. This operation continued all night long until the Commander was convinced he had all the equipment needed, and the ships were cast off to await in the harbor for the fleet. By the time the farce was discovered by the supply officers, that their depot had been looted, it was too late, it was gone; on its way to Island "X."

Most of the men in our battalion boarded ship and sailed out with a convoy of ships, on February 8, 1944, but a detachment of us was left behind; and we did not know the reason why. Then on February 12, our left-behind section boarded a beautiful passenger ocean liner, the "MormacPort," and left Pearl Harbor with only one destroyer accompanying us. Our small Seabee section was traveling with the 22nd Marine Regiment. The ship was crewed by Merchant Seamen (all civilians), and they let us in on a few bits of information. This was the first Merchant Marine ship to be allowed into the combat area. It also, was fast enough to outrun any submarine. We then found out these Marines were to be the ones to invade the island. We didn't quite get the picture of why us Seabees were on the ship with the Marines, but we surmised the reason. Oh, Oh!

We made good time, and for once I didn't get seasick. Every morning we had our submarine alert at dawn, with all hands on deck, in case of a sinking, by torpedo from a Jap sub lying in our path. The third day out we passed the fleet that contained the rest of our Seabee battalion, and we left them behind. After seeing the security of such a large convoy, we really did feel all alone on this vast ocean, plunging into enemy waters, with only one little destroyer to protect us.

The food was good, and most of each day was spent on deck, eating my favorite mixed nuts, and mint chocolate which we could buy in the ship's store. We played chess, and checkers a lot, and found we could average about one hour per chess game, which gave us a good log of passing time. (30 more miles each game).

Finally we approached an atoll. It proved to be Kwajalein, which had been invaded on February 1, just two weeks previously. We viewed our first sight of war's recent devastation, as we anchored in the lagoon off the main island. We still didn't know to which island we were heading.

The following day, we steamed out onto the open ocean again, remaining separate from the Fleet, as before. All that day, and another night we sailed swiftly on. The boatswain's whistle woke us at 5 A.M. for our usual dawn sub alert. But this time everyone was especially tense. These were enemy waters, and in the dawning light, we suddenly saw a bright red flash on the horizon, and then a rumble like thunder. More flashes, and booms. We realized a battle was going on in front of us, just over the horizon.

So, this is what war is all about! After months of training we would soon see war at first hand! In our adventurous spirit, we were anxious to find out what was happening up ahead, but we had an inner conflict that said, "Let's just forget it, and go back home!"

We all remained on deck, with our life jackets on while we steamed ever closer to the conflict. Then the scene emerged on our horizon. It was a Navy ship, and air, bombardment that was taking place. It was to soften the enemy positions, in preparation for invasion.

It had become very quiet around us as we entered a deep, narrow passageway between two islands, into the lagoon of the Atoll. (An "Atoll" is the rim of an extinct volcano, where the lagoon is a body of salt water within the crater. Islands have built up on the rim, made of coral, and coral sand.) This Atoll was almost circular, with a diameter of 40 miles. It consists of 30 small islands, with the largest three being Engebi, Eniwetok, and Parry. These three would be the targets of separate invasion battles.

As we entered the narrow passage we saw that these islands were the typically beautiful tropical islands. About 200 yards off our railing we saw many men bustling among the palm trees. We crowded to the railing to watch; when someone cried out: "They're

Japs!" We all automatically ducked down expecting them to shoot at us, but no shots were fired. So here we are at Island "X"! And the Japs still occupy it! Do you mean *we* are going to have take it from them?

We steamed up the lagoon and dropped anchor with several other ships, near Engebi Island. Some of the Marines debarked, and then we headed back down the lagoon, and dropped anchor about a mile off the main island which we heard was Eniwetok, which was to be our destination. The bombardment of the islands had resumed, and what a fire-works display it was!

Because of all the excitement happening around us, I failed to report for a duty assignment, so for punishment I was put on K. P. duty for a day, scrubbing pots and pans in the scullery. But I considered it was worth it, because I got to see much of the historical action taking place.

The battleships, cruisers, and destroyers were in place behind us, and they opened up with their big guns, firing over our heads into Eniwetok Island, and off to the side into Parry Island. This bombardment went on for 48 hours, by which time I had the worst splitting headache I had ever had, from the noise.

At night it was such a majestic sight to watch the battleships, Pennsylvania, Colorado, and Tennessee fire their three, 16-inch broadsides. The fire from the muzzle of their guns would roll in billowing red and white balls across the water. Then the shells, being white hot, would travel over our heads as we watched, and land on the island with a terrific explosion. Occasionally the shells would hit an ammunition dump, which then erupted in blazing fireworks of shrapnel flying in all directions. The mortar shells being fired over us were something else! They would tumble end over end making a thumping sound, and seemed like they were going to land on us, instead of going over us. We were concerned, because once in awhile a shell would misfire and drop short of its target.

In the early morning, after breakfast I was standing on deck when I heard a flight of planes approaching at right angles to the end of the island. Four dark blue Corsairs (F4U's) appeared, in formation. I would estimate they were at about 3000 feet altitude, just above an elongated fleecy, morning cloud. Just before reaching the island, they peeled off, one after another, and came roaring down the slot directly in front of us, 'til just above tree top level, strafing with their wing guns, and when level let loose their bombs.

What a beautiful sight!

Only beautiful for us on the giving end, I'm afraid. (One of the friendly Christian natives, whom we were to later meet and befriend, was running to hide, when this foursome strafed, and he was struck by a bullet in his arm, breaking it, and knocking him to the ground.

Yet he was so glad we had come to save them from the cruel Japanese. As we conversed with him in Pidgin-English he told how he had become a truly born-again Christian through a Dutch missionary, by the name of Evans. (As a valued souvenir, I have his hymn book from which he learned so many of our old, familiar hymns.)

("Happy," Our Christian Native Friend)

At the time I saw these planes peel-off, I thought to myself that this was the most beautiful war scene I ever hoped to see. It was picture-book in its perfection. The Aircraft carriers covering this invasion were the Sangamon, Chenango, and the Suwanne, on which I was to spend my last three months in service, on a cruise to Okinawa.

As we were told, the air strike was to be the last event to precede the beach-head landing. And so it was. The other ships around us began to disgorge their Army troops, into landing craft. Then like a parade the craft loaded with troops spread out into two sections. One section headed for the Southern, larger tip of the island, while the other headed for the middle. With my binoculars I was able to watch them in detail. They successfully made the beach-head, and clambered up the soft sand to the edge of the trees. From there they disappeared into the jungle-like mess of broken palms.

About four hours later word came to our ship that the Army troops had gotten into trouble and were cut-off by the Japs. The Army had been allowed to cross the island, and then the Japs had cut them off from their beach-head and supplies. The 22nd Marines on our ship were alerted that they were to land and rescue the Army.

We watched them gear up, and then climb down the landing nets on the side of the ship, into "Higgins" landing craft. About 32 men were in each craft, and it was driven by a "Coast-Guard" sailor. As they each loaded, the craft would circle around, waiting for all craft to fill. Then came the signal to head for the beach. With sniper fire keeping all heads below the sides of the craft, they made their way to the beach, landing several abreast. The ramp was lowered and all the Marines ran out, and up the beach.

Reports were received back at our ship as to their progress. The Marines had broken through the Japanese trap, and were sending the Army back to their ships. It was now a Marine operation to take the island. This had been their contingency plan in case the first wave had not been successful. The resistance from the Japanese had been much greater than anticipated, and their forces were more numerous than reports had indicated. It turned out we were fighting the "Imperial Marines," specially selected and trained warriors, each standing half a foot or more taller then the average Jap soldier.

I stationed myself on the upper deck, facing the island, watching the action through my binoculars. This position was well shielded from other deck activity, and especially out of view of the officers. That way I wouldn't be called for other, *interruptive* duty! At chow time I would go down to the galley, and make myself a sandwich, and with a canteen cup of coffee would again take my observation post. I saw action taking place on the island just a few hundred yards away, and yet out of harms way. On February 21, the third day of fighting, I watched our Marines approaching the Northern tip of the island, which had much open, sandy terrain. A Jap pill-box was in a small grove of trees and brush, in the center of this open area. The machine-gun was taking its toll on our troops. I saw one brave soul make a solo dash toward the dugout with a grenade in his hand, but he was gunned down before he got halfway, throwing his grenade toward the enemy. A medical corpsman approached the wounded man from a gully and managed to drag him to a shell hole. The standoff with the Japs continued with firing from each side. I wondered why they didn't use a flame thrower, but I guess the distance was too great.

Soon, as I watched, one of our tanks approached out of the jungle. They evidently had a plan. They positioned the tank on a knoll with the front of the tank pointed downward, directly at the pill-box. In one salvo from the tank gun they hit the dugout point blank. It erupted in flame and sand. Marine troops stormed the box only to find dead bodies.

It was the pause that refreshes. The island was wrested from the Japs. Soon a group of Marines appeared out of the jungle with about nine prisoners who they forced to go into the lagoon to take a bath. Then some of the Marines did the same after the Japs were out.

The next morning all Seabees on our ship were ordered to a lower deck for a briefing by officers. They gave us a rundown on what had taken place, and what was planned yet to come. An officer explained the tactics Marines use in this island campaign. They dig no trenches like the Army does. They move ahead with stand-up fighting, surprising the Japs with their swift progress. Their idea is to kill most Japs in their forward push, but realizing they will over-run some who they do not see in their foxholes, or under palm fronds, in hiding.

Flame throwers, and grenades are used, along with rifles and hand-carried machine guns. They use very few tanks, because of the jungle terrain. Of the thousands of Japs on the island, the Marines realize there are probably hundreds, still in hiding after they sweep through. The Marines figure the island is "Secured," but still requires "Mop-Up" of the remaining enemy forces.

The officer then got to the real point of his speech. "Some of you Seabees have the distinguished honor of being the Marine, 'Mop-Up Squad.' You trained for combat, in the swamps, and woods of Mississippi, and now it's your turn to produce the real thing. Now, Lt. Williamson, will you read the list of 100 men who will participate in this operation. To successfully perform the work our battalion was sent here to do, we must be sure there are no enemies remaining to harass our construction!"

The designated officer climbed halfway up the ladder (stairs), so as to be both seen and heard, and proceeded to read off the names alphabetically. Oh, how I was hoping my name was included! Sure enough, the only Rasmussen in the battalion was called up. With a huge burst of applause we all accepted this challenge.

(Chapel Service, asking God's Safety for the Entire Landing Party)

They announced there would be a brief Chapel service on top deck, to invoke God's blessing upon our mission, and to pray for the safety of every man. Most attended, dressed in our combat coveralls. It was a very sober and memorable service, as we each wondered what the future held for us.

(Heading for the Beachhead)

After the service we each prepared our backpacks, and collected our bandoleers, ammunition belts, grenades, and gas mask, and plenty of ammunition clips, together with our rifles. We were loaded down! Then came the order—"Move em out! Over the side! Into the landing craft!"

So many times we had climbed landing nets during our training, but this time it was the "real thing." A person can never forget the emotion of setting feet in the landing-craft that will deposit him on the beachhead. It's like he's leaving all security behind, and advancing into the totally unknown! Three craft were loaded, and we cast off.

Suddenly: "Duck your head! Snipers!" We kept our heads down, even though we had our steel helmets on. Can you imagine the heat of this tropical scene? It was 120 ° F. in the air. And we wore rubber-impregnated combat coveralls, with a back pack, leggings, socks, boots and helmet. But our main thoughts weren't riveted on comfort; rather: "What do we do next?" We were instructed to use the buddy system. Always two of us together; one to cover our backside while the other foraged for Japs.

Beware of booby-traps. Take no souvenirs! Yet, Seabees were noted for finding some of the best keepsakes.

The tide was low, and numerous coral clumps blocked our approach to the beach, but the coxswains maneuvered the craft skillfully, and at full speed rammed the "Higgins Boat" onto the beach. The front ramp dropped, and because I was being pushed from those behind, I was one of the first to jump off into the water which was knee deep. The beach was steep, up to the jungle edge, but we made it up fast. We stopped at the jungle opening, shocked at what we beheld!

A column of Marines was just returning from the front lines, and we stared in utter disbelief! These guys had been with us on the ship, and had just survived three whole days and sleepless nights in exhaustive combat. Even their own mothers wouldn't have recognized them. Staggering, exhausted, flame-thrower soot covered, haggard, some bandaged. My heart went out to them! What a pitiful sight! All that they had suffered for our loved ones back home, and yet these were only the ones who survived. I can't adequately describe what a young man goes through in horrible combat, often hand-to-hand with bayonet, or knife, in oppressive heat, death on all sides, shooting fellow-men; kill or be killed!

The thought often crosses his mind: What in the world am I doing out here. My enemy and I don't have anything against one another individually. Yet we are both fighting for our country and its ideology. Oh, the folly of mankind! Yet each must fight on, to protect what they believe in. This scene of battle-hardened, exhausted fighting men was what confronted us first. It left an indelible impression upon us that will remain as long as we live.

Flies were swarming everywhere, attracted and breeding on the sweet sap oozing from broken palm trees and branches; but worst of all, they were feeding on dead corpses also. This was a dreadful potential for disease. We gathered in a clearing and were told we could stow our back-packs here, to be picked up in the evening. They promised us we would have a water tank truck available at noon for re-supplying our water canteens. We each had filled our canteens before leaving the ship, and we were urged to drink often.

We chose our buddy to accompany us on our hunt for Japs and we took off. We were to cross the island and continue bearing South, criss-crossing the island. Shoot to kill! Don't turn a body over because he might be lying on a live grenade. Blow up bunkers with grenades. Report any dead Americans you find. Dead Japs will be picked up later, by the "Sanitation Squad." You are here to finish off any live ones. Expect suicide attacks. Check out anything that looks suspicious, or is a likely hiding place.

It would be a long, hot day as my buddy and I progressed. We lost track of the other Seabees as we penetrated the jungle of palm trees, broken fronds, and underbrush. The stench of burned flesh and palms was overwhelming. An indescribable combination! Flame throwers had been used extensively, which set fire to anything it touched. Damp as it was, the smoke was still rising from smoldering stumps and palm fronds on the ground.

We emerged into a glade of grass, and were greeted with a scene which before military training would have caused me to lose my breakfast. In a circle of about 100 feet there were about 15 bodies, lying where they had fallen in hand-to-hand combat.

(Aftermath Scene of Hand to Hand Combat)

I approached one that was lying on his back and had a bulging pocket. I thought, maybe he had some documents that would be useful to our intelligence unit. I took out my battle knife and slit the pocket open, taking out the papers. As I looked at them, I discovered they were all in English, and a picture of a Negro man and his wife. Oh, No! I looked at his face again, and sure enough this was an American black man, hardly recognizable because of soot and grease. (A dead Japanese, and Americans often looked alike because of dirt of battle, and their uniforms were equally confusing.)

I stuffed the papers back into his pocket and called to my buddy that we should get out of there, fast. You can be shot on the spot for looting the body of our own men. And the stench was terrible!

We continued on until we arrived at the ocean side, and there we found several tank-trap fortifications, with a number of dead Japs sprawled in the ditches, along with twisted steel. They had evidently been blown up, as the Marines made their sweep. Nothing was left alive here as we checked each body.

Turning back into the island in a Southern direction we came to another clearing. There was a Jap tank with its treads broken. It was my turn to check out the territory, so with my buddy holding his rifle at readiness, I climbed onto the tank, to look into the open turret hatch.

(A Japanese Tank Disabled in Battle—a Hideout)

It stunk awful! As I looked in, I saw nothing but blackness inside. I turned around and said: "No one is alive in there, they couldn't survive the smell!" And jumped down off the track. We walked across the clearing toward a bunker we saw, intending to check it out. Shouts behind us made us turn quickly, just in time to see a Jap climb out of the tank hatch, jump down, and start running for the jungle. The Marines who had shouted, opened fire on him, and he went down. (Another "good" Jap) as we thought at that time. The Marines congratulated me for flushing him out, but they thought I was lucky for not being killed myself. I knew the Lord had His angels watching over me;—I thanked Him!

We continued on our way to the Jap bunker; the first of its kind we had encountered. It was dug into the sand, and covered with coconut logs. I volunteered to check it out, so going down the ramp entrance, and holding my rifle at readiness, I looked into the tunnel that was at right angles to the entrance. Again, it was so dark I couldn't see a thing in there.

We disobeyed orders by neglecting to blow it up with a grenade! But later in the day we came by there again, and someone else had blown it up. The tunnel was shaped like a "U," and at the blind end of the bunker were three freshly killed Japs, killed by the grenade that blew it up. What a "Dud" I was becoming on Mop-up.

We made our way down toward the Southern tip of the island, viewing battle scenes most of the way. Then we heard machine-gun fire, and shouting ahead. As we approached a group of Marines standing by a bull-dozed trench, we realized we had come upon one of those reverse "atrocities." Several Jap prisoners had been brought in, and while standing in a column, a Marine, just returning from battle on the Northern front lines, had shouted curses and mowed the Japs down with his machine-gun. We later heard he was court-martialed, but found to have mentally "cracked-up," and was sent to a treatment center. War is horrible!

We came across a concrete block-house where the Japs had kept the Christian, Dutch missionaries in captivity when our fleet arrived. Our Marines had rescued them and then destroyed the block house. By noon our canteens were empty of water, so we had made our way back to where we had left the other Seabees. We were all out of drinking water, and were told to wait until the water tank arrived. It was finally unloaded from a barge and we lined up to fill our canteens. A shout went up! Don't drink the water; the tank was used to haul creosote, and the water is contaminated. Being so thirsty, I decided to take a taste of the water. It tasted like tar. Then the word came that no water was forthcoming in the foreseeable future. Well, we had to do something to quench our thirst, so we cracked open coconuts, and drank the juice. It tasted good, and quenched our thirst, but very soon the effects had us all digging holes in the sand to relieve our diarrhea. What a time! We learned it was an instant cure for constipation.

For the remainder of the day we were assigned to unloading supplies from the barges that were bringing them in from the cargo ships. Then came the surprise of my life! We were working with a bunch of Marines who had just arrived from doing battle on Guadalcanal.

Looking for the restroom on the large barge, I clambered over the cargo to go down into the hold. A battle-hardened Marine was climbing up the ladder, and we met. He looked so familiar, and I stared at him so intently he growled: "What's up with you, Mac?" "Are you Johnny Yarmonchik?" I asked. "What's it to ya?" He snapped back. "Do you know who I am, John?"

He stared at me a moment, then shouted: "Clayton!" Here was a boy I hadn't seen for seven years, when we had grown up together in Pennsylvania. We had been the best of friends in grade school! What a reunion that was! Seven thousand miles from where we had last met. We talked until we knew we would be in trouble to stay longer, and each went back to his duties.

Because no one was doing any cooking at this time, we were scheduled to eat K-Rations for 12 days. These consist of a package of prepared food about the size of a pound of butter. It had a small can of main dish, which was usually good; three hard dog biscuits; a bar of tropical chocolate; a packet of instant coffee or coolade; and 3 cigarettes. To say the least, they weren't very appetizing, but they did somewhat fill a void in our stomachs. Except the cigarettes, which I threw away. After 12 days they were required to give us at least C-Rations.

It so happened that, as clever Seabees would have it, a box of C-Rations fell out of the cargo net from a great height and broke open all the gallon cans. Not wanting the contents to spoil we all took a can; and Johnny and I sat down on a log and devoured an entire can. C-Rations are like corned beef hash, and this first time eating them was like having a banquet! As the 23rd Psalm says: "Thou hast prepared a table before me in the presence of mine enemies." And so it seemed!

(Our First Night on the Beach,
Lt. Commander McAllister, my Neighbor)

As the day drew to a close we were told to pick up our gear and retire to the far side of the island and set up our Pup-tent on the beach overlooking the ocean-side reef. My buddy and I put our shelter halves together, forming our Pup-tent and had a select spot on the beach, far enough from the jungle to make it difficult for any Jap to sneak up on us. Three feet from us was the Pup-tent of Lt. Commander McAllister, our battalion second in command. Being that we had an ideal camp spot, other officers camped near us. After supper of K-Rations our water tanker arrived and we were able to fill our canteens with real water. What a relief! We could now also boil some water for making instant coffee.

As darkness took over we climbed into our bedroll and tried to go to sleep, with waves crashing over the reef, and in trying to fit our body shape into the sand. Then suddenly the sky became as bright as day, and machine-gun fire opened up about 300 yards down the beach. Japs were coming out of hiding and trying to make their escape over the reef to the neighboring island, Parry. The Marines were on the alert, and in the light of the flares, few could escape their gun-fire. This process continued all night, with fitful sleep resulting.

The next few days we all became quite adept at unloading cargo. Our expert crane operator would pick up a cargo-net load from the hold of the barge, lift it high, swing it over our heads, and swiftly lower it to an exact spot on the beach. From there, we would each take a personal load of cargo and deposit it for waiting trucks about 100 feet away. The hardest of all cargo proved to be the 100 lb. kegs of nails. We would put a keg on our shoulder, struggle up the beach slope, slipping and sliding in the loose sand, till we dropped them at the top of the slope, exhausted. But no let-up, lots more to go.

Then came the welcomed call from Chief Barger, in charge of the Machine-Shop trailer. "Rasmussen, get down to the barge unloading our trailer. We are to set it up, and get to work." I had been waiting for that call. I was beginning to think that maybe my selection for the shop crew was being changed, because of all that had taken place from Hawaii to here. I had seemed to have different priorities, assigned to me. However, by March 8, 1944, we had the shop in operation, and I was one of the regular crew. There were a large number of machinists in our battalion, and some of them let it be known they were upset at not being selected for the shop crew. But the Lord had brought about my assignment, along with only six others, all of whom had Petty-Officer ratings, with me being the only Seaman rating.

The officers, who were my Pup-tent neighbors, began to think it wasn't so desirable having an enlisted man living in "Officers Country," as this was now called. So I was politely requested to find sleeping quarters in one of the 6-man tent cabins being erected.

In response, I found our Company officer and was assigned to a tent a short distance from our shop. It was beginning to seem a bit more civilized. Tent cities were going up all over the place, using the pre-fabbed pallets our carpenters had built in Hawaii.

Then one day, after sixteen days with no mail we heard the wonderful words: "Mail Call!" What excitement! You can't imagine how important it was to get mail from home. And I wasn't disappointed! Letters from my parents, my sister, my brother, and the one I was most curious about, from Cora Jacobson. I had requested pictures of her in my last letter. Was she interested enough after seeing my picture, to send me one of herself? "Great Day!" She included three pictures: one a picture of herself in a graduation formal; one in street clothes on the way to summer camp; and one in shorts with a bouquet of wild flowers just picked in their Northwoods. What a Gal!

(North Woods Flower Picking)

(High School Graduation)

(Off to Camp with Sister Alice)

These Three Pictures Were the First Ones
Received from Cora Jacobson

This was only the beginning of many exciting happenings on the three-mile long island of Eniwetok, where we spent a total of seven months. Almost every day my friend Johnny and I had a chance to visit, but all else must wait to be disclosed at another time.

In all God's perfect wisdom, He chooses to disclose only one step at a time of His plan for our lives, in bringing us into conformity with His Perfect Being. These steps of experience have taught me in varying degrees what God meant when He said:

> "For this reason, make every effort to add to your faith goodness; and to goodness knowledge; and to knowledge self-control; and to self-control, perseverance; and to perseverance, godliness; and to godliness, brotherly kindness; and to brotherly kindness, love. For if you posses these qualities in increasing measure, they will keep you from being ineffective, and unproductive in your knowledge of our Lord Jesus Christ." II Peter 1:5–8.

I would speculate, that is why it takes most of us a life-time to conform.

—A VET

Air-Raid
Bombs Away!
Japs Attack
Utter Terror
No Athiests in Foxholes

Chapter 3

At Target "Zap" or (Saved By a "Dud")

Along our pathway of life we occasionally experience an event like none other in our past. It's new; and it produces emotions that have never before been awakened. At such a time it is comforting to have a loving, omniscient companion, the Lord Jesus Christ who is a comforter, a shield and a shelter, in which we can hide.

—A VET

It was quite a pleasant transition to move from our 2-man pup tent on the beach of Eniwetok, to a 6-man tent cabin. We had a tent that was made up of a raised wooden floor, with a framework of 2x4 walls and rafters. This structure was covered with water-proof canvas, with screening on all sides for air cooling, with roll-down sides to keep out rain. It was even equipped with a screen door. In the center of our hut was a table built around the center-pole. We each were provided with an army style folding cot. Any other furnishings would need to be provided by each with his own ingenuity.

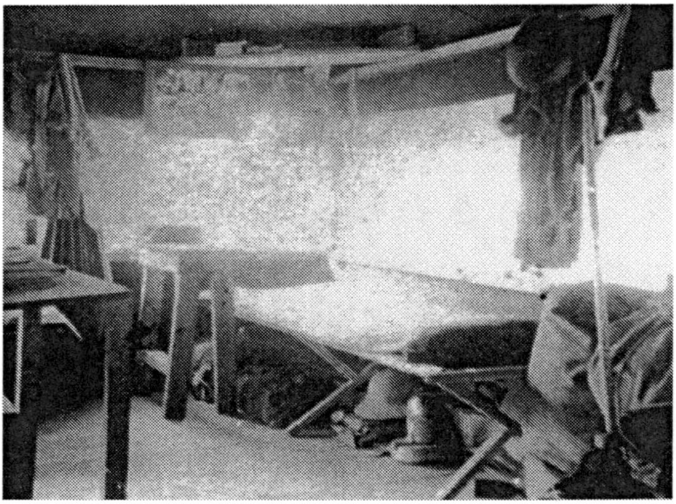

(My Section of Our 6-man Tent)

My tent mates were: Ganske; Kennedy; Smith; Trigg; and a later mate, Walters. These fellows, including myself would make up our tent roster for the remainder of our time overseas, both on Eniwetok, and on Tinian. We all got along well, even though we proved to have very individually distinctive personalities. Each worked in different job functions. Carpenter, truck driver, grease monkey, welder, office machine repairman, and machinist. I was the adventurer, but was usually successful in getting one or more others involved in my explorations, or more unconventional activities. I took many pictures, and wrote letters to many more people than the others. Consequently I built a writing desk. My correspondents included sixteen young ladies, from California, Pennsylvania, Illinois, Wisconsin, New Jersey, Arizona, and Wyoming. All but three of them were friends from places I had previously lived.

An interesting incident occurred because of my photography and letter writing combination. I was writing to two Pennsylvania girls that I did not realize knew each other, and they both happened to attend the same community gathering. I had sent each of them identical pictures of me, but had color tinted one of them, and not the other. Well, they showed each other a picture of their military boy friend, and who should it be but me. The girl with the plain black and white picture was suddenly very jealous. (I would never have expected it of her, in particular). I received a letter from her that indicated her emotions, with a kindly request for a *tinted* picture of me. To say the least, I complied, with a new appropriate one with her sailor in a striking pose with a surfboard, finished in the best tinting I could muster. It did the trick! But now a picture feud was underway by two very smartly beauties. What fun I had; all in an innocent way, that kept both of them as friends!

Incidentally,—my parents had gifted me with a "Marshall's Photo-tinting Set," which provided many enjoyable hours of recreation time, when combined with a photography hobby. Before going into the service I had bought a very good camera. It was sent to me overseas after our Eniwetok invasion. Of course, these were the days before color photography, and to have pictures in color required painstakingly painting them with transparent tinting. During wartime, film of any kind was almost impossible to buy, so if we wanted a pictorial record of any kind, it was necessary to procure the film. My personal method of obtaining film will be revealed in a later account.

Work had begun on building our new airstrip, and to obtain coral with which to pave it, they blasted the coral all along our beach, out to about 100 yards. This operation went on day and night. The familiar cry just before a blast was: "Fire in the hole!"

We had to be constantly on the alert for such a cry, because the blast would often send large chunks of coral flying into the air. Once I was riding in a jeep with three other guys when a blast occurred right next to where we were driving. The driver immediately hit the foot brake without putting on the parking brakes. We each jumped from the jeep, taking refuge alongside, and under it, while the jeep was still rolling, with us trying to stay protected from the flying coral, while keeping from being run over by our own vehicle. It was laughable afterwards.

(Artist's Conception of: "Fire in the hole!")

However one night, while we were all sleeping, they were blasting only a couple hundred feet from our tent, when we were awakened by the cry: **"Fire in the hole!"** We all baled out of bed, and under the center table. **Wham!** A large chunk of coral came ripping through our tent roof, and plunged right through my cot. Of course, I had a stitching job to do on my cot canvas the next day, thankful it wasn't the doctor that had to stitch me up.

For the first couple of weeks, we had a rather crude soup-kitchen setup for serving chow. The cooks had the first of the Quonset huts for preparing the food, but we collected our food in an outside line, and then sat around on coconut logs to eat it out of our mess kits.

They had furnished us with face covering mosquito netting to protect us from the mass of flies which were present everywhere. As we ate we would raise the netting to take a bite of food. No matter how careful, or fast we were, we always ended up having to spit out half a dozen or so flies, before chewing our food. No kidding! The possibility of disease was constantly present!

It didn't take long before the dysentery plague was in full force. It started with just a few, and by the hour it gathered speed. Soon the dispensary tent was full with sad sack Seabees, and they had to start using other tents. They ran out of space, so cots were set up outside. There was a steady stream of sick men on the way to the slit trenches dug to relieve the diarrhea. Some of the men could not make it to the trench before they had to make an emergency stop along the way. Oh! What a sorry looking sight! Before the plague ended only a few of us didn't get it. The cause was flies that were feeding on dead flesh, and then getting in our mouths. The sanitation squad was called into overtime to spray everything in our proximity. They later were able to spray the entire island, and to rid us of the plague of flies. Praise the Lord!

The transition of food supply was like this: The first 12 days we ate K-rations. Then we progressed to C-rations, and in a few days we started to get real food. By the end of the first month, our Seabee galley was the talk of the island because of the good food being served. We had overcrowding with Marines, Air Force, and Navy, which required that we get into the chow line early in order to get the regular food.

(Seabee Chow Time Crowding)

One Sunday they served chicken and all the trimmings, and ran out of food because of so many outsiders going through. Our officers explained we were being served such good food because our island was the closest to the Japanese homeland of any allied holdings. They expected the Seabees to build the airstrip that would bomb out the Japanese held islands, so they treated us special! Sorry, this food situation didn't last very long, because soon we were back on C-rations for every meal, along with other staples. Our cooks tried their best to prepare it in a different way for each meal, but it was still the same C-rations even though called breakfast ham and hash browns, lunch corned beef hash, or dinner potato/meat pancakes.

By March 8, 1944, we had the Machine Shop trailer set up, and were busily machining, on twelve hour shifts. At first most of our jobs were machine repairs on our rolling stock, jeeps and trucks. I ground valves, and made clutch forks till I didn't want to see another one. I became the clutch fork specialist, because I was the most efficient one with a file; tiring, but we had to keep the vehicles moving!

I didn't have time to dig a foxhole by our tent, so I built a special one near the shop. We had a pile of pipes stored behind the shop, so I chose to dig my hole just at the edge of the pipes, but underneath them. My calculations were that even if I took a direct hit by a bomb, the pipes would roll over my foxhole, and cover me from shrapnel.

(Our Eniwetok Machine Shop)

Now I was ready; and if an air-raid took place at night I would be right near my shelter, since I was working most of the night. Whereas if a raid took place in daylight, which was not likely, I probably would still be near enough to the shop to run to my fox-hole there. Since we still didn't have our airstrip finished, we couldn't send out reconnaissance planes to discover the Japanese strips, for bombing. So we were quite sure the enemy bombers would find us. And it wasn't very comforting to know that because our island was so small, we would all be in great danger of being hit no matter where we were.

(My Foxhole, Under the Lower Pile of Pipes)

The Marines set up an anti-aircraft battery just a short distance from our shop, being very careful to level it in the sand. We were puzzled, because wouldn't the sand shift under the recoil of firing? It was an interesting operation to watch. Would the enemy try to bomb that battery out first, and maybe hit our shop instead? My imagination was running at high RPM, so I dug my foxhole a little deeper!

Meanwhile I found where my Marine buddy Johnny Yarmonchik was housed, and looked him up. We had such a great time catching up on the news from back home, and hearing what each had been up to since we had last been together seven years ago in Pennsylvania.

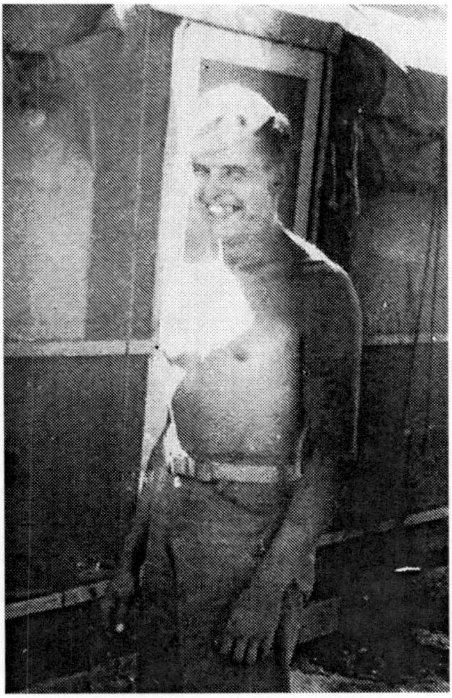

(Marine Johnny Yarmonchek)

(Clay, on Leisure Time)

He had news received by recent letters from our old grade school teacher, Miss Weaver. She had told him what all our mutual friends were doing. Johnny had known my parents well, and was glad to catch up on their latest. He had just come from the battling on Guadacanal, and had been out in the Pacific theatre almost four years. His specialist job was as a meteorologist. Each day he would launch a weather balloon, and track it for winds aloft, to relay this information to our airbases so they would know what to expect over our island area. The Marines were already back to full military regulations, despite such recent battles. However, he was free part of every day to come and visit with me. We had such good friendly times together.

The early morning hours of March 8, only two weeks after our invasion of the island, will always live in my memory! A fellow machinist, Derflinger, and I were on duty in the machine shop, when the first air-raid siren went off at 4 A.M. Immediately we doused the lights, and waited near our foxholes. The Marine anti-aircraft position was buzzing with activity. We tried to hear what they were saying, to get information on the action. The Japs were attacking Engebi Island 40 miles North, at the far end of the lagoon.

Soon their searchlights went on and they focused on two enemy planes at about 15,000 feet right above us. The lights must have blinded the pilots, as they just kept flying straight ahead. The Marines opened fire with the ack-ack gun, and we saw the shell burst a ways behind the planes. Each time the gun fired they missed by a much wider margin. Soon they were firing about two miles behind the planes. It turned out the gun pods were sinking into the sand and the gun was way out of level. The radar controls were unable to coordinate the track of the planes accurately, with the shifting gun position. We felt terribly disgusted, and upset at such lack of defense!

As the planes disappeared into the distance the searchlights went out, and all became very quiet. We stood and watched as Jap planes continued to bomb the other island and many fires and explosions were lighting up the sky. As we watched the action, but hearing only distant rumbling of explosions, we started hearing a strange whistling sound gradually getting louder. It became very shrill, and we discussed momentarily what it might be. Then we heard a new swishing sound, and it dawned on each of us simultaneously that it was a bomb coming in our direction from a dive-bomber, with engine shut down. We dove for our foxholes, and I entered mine lengthwise, banging my steel helmet on the pipes as I slid in! Putting a big dent in it, and glad it wasn't my head.

Wham!! The first bomb exploded up the island a ways from us. The sand on the sides of my foxhole was shaking down, and I started digging with my bare hands to make the hole bigger, wiggling down into the bottom sand! What a fright! I'd never been so scared in all my life! Realizing they were trying to kill us!

WHAM!!! The second bomb exploded, this time it was much closer! Oh, why didn't I dig my foxhole deeper? I'll dig some more right now! More sand was falling in on me. Wow, the next one will be right on top of us! No time to do anything now except pray! And I did! It didn't take long, just a short one: "**Lord, help me**!"

THUD!! What was that? Sand flew everywhere; pinging on my pipes, trailer, barrels! But no explosion! Then at a greater distance on the reef, another explosion, **Boom!!** Then silence! But keep down, there may be more! Then Derflinger called out: "Ras, are you all right?" Boy it was so good to hear another voice. We decided to stay in our holes for awhile longer, to play it safe. Then we ventured out. The first thing my buddy said was: "Now I know why they say; 'there are no atheists in foxholes.'" We wondered where the "dud" bomb was, but we were not about to explore. Soon the all clear siren sounded, and we could turn the lights on. Shouts up the beach a short ways indicated a bomb was still there, unexploded. It was right in the middle of our dynamite and ammunition dump. The bomb disposal people were called, and they defused it.

God had answered my hasty prayer! He had kept the bomb from exploding! If it had gone off, the entire explosives dump would have blown up, most likely taking all of us with it. But it's potential devastation was divinely limited to only spraying sand on a bunch of us!

You can be assured we all became very conscious of the need for foxholes. Mine was dug deeper, and farther under the pipe stack; and I also dug one by our tent.

This particular bombing was the most frightening experience of all the war for me; including the dreadful anticipation, and again the actual bombing. It's amazing that such a vast amount of frightening experience can be crammed into only a few seconds of time.

We now prepared to experience many more such wartime occurrences, and it gave all of us more incentive to get our own airstrip built. Then we could send our own bombers out to eliminate the Japanese air bases from which their planes were coming. It was a deadly match, with our lives at stake!

> Surely the Lord will save you from the fowler's snare and from the deadly pestilence. He will cover you with his feathers, and under his wings you will find refuge.—You will not fear the terror of night, nor the arrow that flies by day. (Ps. 91:3–4)

It was very comforting to read this in God's Word, at that time, and to put my continual trust in Him.

—A VET

(Seabees Setting up Our Tent Structures)

POW

DIVINELY DIRECTED DELAYS

AMMUNITION BARGE BLOWS UP!

GALLEY DESTROYED!

MOTHER PRAYS!

Chapter 4

Divinely Directed Delays or (Prayer Changes Things)

(Ammunition Barge Explodes)

"Prayer spans both space and time—to the very Throne of God; and His answer knows no limit; extending even thousands of miles from the requester to the recipient.

—A VET

Divinely Directed Delays

As I reported for duty at midnight, the tropical evening had cooled from the daily 120 degrees F., to a more comfortable 100 degrees. The workload in the Machine Shop, where I worked as a shop machinist was extremely heavy. During the night hours the work was mostly performed in the shop, but at dawn the entire battalion was starting their work of building an airstrip on the little island of Eniwetok. Our tasks then took on a variety of responsibilities.

As these daylight projects were assigned at dawn, I always looked forward to whatever new challenge would be presented. In the adventurous spirit of a 22-year old, I welcomed this day! Ever since the day of our invasion beachhead one month previously, I knew each day could be an "adventure," and even might be thought of as dangerous to some—but not to me. YOUTH! What confidence; what daring; or what brash thinking! It was 5:30 A.M. on Eniwetok, March 23, 1944.

Back home in California, my Mother suddenly awoke from an afternoon nap, and was jolted upright in bed! She was horrified! She had awakened from an extremely vivid nightmare in which she saw her son in a hospital bed, pitifully and critically wounded! Her hearts' cry went out to God. She prayed! "Spare my son; my baby, from this tragedy, Lord!"—After reviewing with the Lord her fondest memories concerning me, her youngest son, she committed the keeping of my spirit, soul, and body to Him. Then, confident of answered prayer, she again fell asleep.

Meanwhile, at dawn on the island, the Chief handed out the days' assignments, to be completed before our shift ended at noon. He knew I liked challenges. He had a real one for me this day! "Our workload is too heavy to send the regular two-man team, but Ras, you are to go on the Galley inspection tour by yourself. Keep it moving and finish the three galleys and be back here by 10:00. You can't start 'til after breakfast when the kitchens are clean."

The schedule called for 8:30 to 9:00 at the CB galley, 9:00 to 9:30 at the Marines' galley, and 9:30 to 10:00 at the Air Force kitchens. The schedule was therefore unknowingly set to coincide with a tragic event yet to unfold.—If all was to go on schedule, I would never again see the break of day.—However; *prayer changes things!* After breakfast, I hopped in our Jeep and tore off down the sandy, bomb-gutted road to our Construction Battalion galley.

(Road to Seabee Galley)

Never having done this before, I asked some guidance from the Chief Cook, and he said, "Before you spend your time on other machines, we have this potato masher that's not working. We need it badly for making lunch. Rest assured, I'm not going to let you leave until it's fixed!"

"But Chief, I have to complete my rounds to all three galleys before 10 o'clock, so I just can't spend the time now. Suppose I send someone to fix it later."

I sensed the Chief wasn't about to budge. My mind was racing over the problems: the urgency of having the machine working for lunch preparations; the schedule I must keep, and it was already almost 9 A.M. and I had allotted 1/2 hour at each galley. My next stop was down the beach about a mile, at the Marines' galley, and I was due there at 9:00. And last but not least was my seemingly very significant problem, the fact I didn't know how to fix this broken machine! Delays! Delays! How in the name of common sense could I make my schedule?

However, unexpected things can happen. When I turned on the machine, the problem was immediately apparent, and I saw what needed fixing. However, this meant I assuredly must now stay and fix it! With all-out speed I rushed to the Jeep for my tools and immediately started taking the part off the machine, so I could re-adjust it. The big galley clock stood at 0910, Navy time, that fateful morning.

SUDDENLY,—without warning, the tool flew out of my hand! Simultaneously, the Quonset hut seemed to lift off its foundations with a—*CLANG !* as though I were inside a steel drum that was being beaten. The ground shook! A terrific blast hit the building with its deafening roar and jarring concussion.

My instinctive thought was that we had suffered a direct hit by a bomb! The half-dozen of us inside the building burst through the screen doors to escape outside, and to see what had happened.

Down the beach about a mile a mushroom-shaped cloud was rising, already hundreds of feet in the air, with bright red, billowing balls of flame towering ever higher. It was the most frightening sight I had ever seen!

(My View of Ammunition Barge Explosion)

After gathering my composure, and ruffled nerves, I went back to complete my job, and shakingly finished repairing the mixer. Promising the Chief I'd be back the next day, I left for the shop which was closer to the blast area, noting the fire was still burning, with new explosions sending up more clouds of smoke and flames.

At the shop, as I jumped from the Jeep, I saw the most startled look in the eyes of my buddies, and even a tear being brushed from the eyes of Chief Ward. "Boy Ras! Are we ever glad to see you! We thought you were at the Marine's galley, on schedule. We just got word the galley was completely wiped out! Leveled to the ground, with no survivors! The ammunition barge that blew was at the dock right in front of the galley."

As the day wore on we received further word on the tragedy. Our Seabee Fire Chief Stewart was killed, ten Marines from the galley were killed. Ten or more seriously wounded were flown to Hawaii for surgery, having lost legs, arms, and receiving other very severe injuries. Those who were on the barge could not even be found, and one who was seen diving off the barge just before it exploded was drowned in the water.

It was two weeks later I received a letter from my Mother, relating the experience of her dream. Then I knew why I had been delayed at the Seabee Galley! Due to censorship of all our outgoing mail we were never able to relate to our correspondents any information concerning events of war, such as we had just experienced. Consequently our parents did not know of our whereabouts, or circumstances. The common bond across the miles was that of prayer. God knew both ends of this communication.

I calculated the time Mother had prayed for me. Considering the difference in time zones, my calculations put praise in my heart for Gods' Grace. Mother had prayed for me at the very time I had been delayed at the Seabee galley, just shortly before the barge blew up.

No wonder I had been "delayed" at the first galley; for without Mothers' intervention by prayer, I would have been in the Marine's galley where no one survived! God had answered her prayer of faith!

I can confidently say with the Psalmist David:

> "If you make the Most High your dwelling—even the Lord who is my refuge—then no harm will befall you, no disaster will come near your tent."

—A VET

(Marine Galley Demolished)

HEY SARGE!
WHERE'S THE BARGE?
?

Chapter 5

"Hey, Sarge: Where's the Barge?"

Sometimes we don't tell everything we know. At the time there may be something we're afraid of, or maybe ashamed of. But as my mother used to say to me so often, as needed: "Be sure your sin will find you out." It's about time, I'm thinking, to admit all, and get it off my chest. Finally the secret incident with the barge will finally be cleared up.

—A VET

On June 1, 1944, word came from Headquarters that our 110th Seabees had completed our assignment of building a 6500-foot airstrip. It was known as the "land-based aircraft carrier." We were all very proud of the immense task we had finished, and it was effectively being put to use by Navy B-24, (4-engine bombers), Billy Mitchel B-25, (twin-engine bombers), Hellcat F6F's, Wildcat F4F's, and Corsair F4U's (fighter planes).

(The Airstrip, 52 Years Later)

At times the place was much busier than the largest airport today during rush periods. Fighters took off in formation; bombers loaded with mass destruction departed within seconds of each other, and all aircrews were busy scurrying around getting others ready.

When an enemy attack was detected, the fighters "scrambled" into the air, banking at right angles to the strip before half the strip length was used and at only treetop level. It was all very exciting!

Then one day the chilling word came, along with air-raid sirens, that our island was being put on "Red Alert"! The Japanese fleet, with aircraft carriers, battleships, cruisers, destroyers, and troop ships were heading our way, and an invasion of our island by the enemy was anticipated! We were told to keep our rifles, bayonets, and sidearms with us at all times. Crews were dispatched to place rolls of barbed-wire barriers in three echelons on the ocean-side reef. Anti-aircraft batteries were on ready alert, and all our machine-gunners were issued ammunition. It was preparation for an all out battle. We were rightly scared!

Gone was our complacency. The thought of facing the enemy storming our beaches was so frightening we hardly dared think it! We waited with much fear and trepidation. The thought of unknown dangers that lay ahead was almost petrifying. The entrances to the lagoon were guarded with big guns, so we were sure the enemy would have to attack from the ocean side of the island. At such a time it is hard to think of much else than the events that might lie ahead. We waited. Night came. Morning came, and still no word.

Then the news flashed that our own fleet had intercepted the Japanese fleet and had what became known as, the Mariana's "Turkey Shoot." We had blasted the enemy ships, and shot down over 300 of their carrier planes. Whew! What a relief!

(U.S.S. Lexington Pilots After Return from "Turkey Shoot")

These pilots intercepted a squadron of 20 Jap Zero's and shot down 17, with no loses themselves, so they were celebrating!

My prayers in desperation had been answered. God knew the end from the beginning. I should have had peace in Him. Again I had failed to have the faith and trust that I had been taught He could give. Sometimes our experiences teach us more than a million words could provide, when we have God's Word as a background. We are to live, and grow spiritually, as He leads us. He knows far better than we, what it takes to provide us spiritual maturity, and I was learning.

It was easy to draw a parallel of our preparation for an invasion, with the promised return of our Savior, Jesus Christ, when He comes to catch away His believers from this earth for a glorious home in Heaven. For God tells us: "In such an hour as you think not, the Son of Man will come." He bids us to be ready, for at any time He may come.

We began to settle down from the latest threat of the enemy, and gladly noticed we were no longer experiencing night air raids either. This was a great relief! Our senior officers notified us we had been given a further battalion assignment to *maintain* the island's facilities until we left for our next invasion, on Tinian, in the Mariana's Islands. This provided our shop enough time to design, and build a "Compass Rose" in the parking area of the airstrip. This device was built from a modified Japanese tank turret, and buried flush with the surface of the strip. They would roll an airplane into place, with one wheel on the turret, and pivot the plane around a precisely segmented 360 degree circle, and were thus able to calibrate their plane's compass.

(A B-24 Prepares to Use the Compass Rose)

They then had a chart showing the deviation of their compass from that of true directions. They had a critical need to navigate using their compass, in order to find their island base, while flying over hundreds of miles of unmarked ocean, always in radio silence. At the time, this facility was the only "Compass Rose" west of Hawaii, so planes flew to our island from other bases to calibrate their compasses. It was quite an accomplishment for our little shop to facilitate such an important function for our aviation fleet.

As a result of its importance, one day a jeep sped up to the shop, with a lone officer driving it. He jumped out, and came up to me, who happened to be the first and only enlisted man in his path. He addressed me in a commanding voice: "Come with me, sailor, I want you to show me the operation of the 'Compass Rose,' and explain to me how you were able to build it in your shop, using only salvaged Jap material." This order caught me by surprise, so I looked around at my Chief, ready to ask him if it was all right to go with this officer. He motioned me in a convincing way to go along, without question. Never having seen this officer before I did not realize he was the Captain in charge of the entire Eniwetok Atoll. He was the distinguished Captain Cruz!

(Our Jeep Pulls Out, Heading for the Airstrip)

He asked me to drive us to the site, where I proceeded to explain how we had modified the Jap tank turret, machining it to the precision tolerances necessary to do such an accurate job. He was amazed, and expressed his pleasure at our innovation, and said we were to be commended, for such an important contribution to the ongoing war effort. On our way back to the shop we had a more relaxed conversation, as he asked personal questions about my home, and family. We left as friends, and I was quite proud to have had such exposure to the top brass of our Navy Operations. The Lord was so good to me throughout my service days in providing friends in both the enlisted and officer ranks.

It was at this time that a large U.S. fleet began to assemble in our lagoon, in preparation for the invasion of Saipan, the first island to be invaded in the Marianas. We had completed building a main road completely encircling the island, and traffic was quite heavy, as the activities were increased. We even had an occasional traffic jam.

In modernizing our facilities, we had gone to more sophisticated bathroom accommodations, similar to the outhouses used in rural communities, in the States at that time.

(Artist's Sketch of our Previous Facilities)

(The reader may not be aware that back home, houses in the smaller towns did not even have running water in the house, and of course did not have indoor bathrooms. Outhouses were usually in the back yard, for toilet facilities).

Well, on our island outhouse buildings had been constructed, usually alongside the road, for convenience. Because we were an all male community, there was no need to provide any sidings on the buildings, so open screening was placed around the entire framework, to allow the trade winds to cool and to carry away the odor. Unfortunately, it happened that I was seated in one of these 10-holed outhouses next to the road when a jeep load of sightseeing female nurses came cruising by. They were stopped by a traffic jam just ten feet in front of me. Giggles made me very much aware I was on display, so I made as though this sort of thing happened every day, and tried to ignore the situation. What an embarrassing situation! I hadn't seen a woman in 118 days, and now I didn't even dare look!

While trying to keep cool from the excessive heat, our passion for swimming continued to flourish, and now we needed more excitement, so we built a diving platform with a springboard. From this we all experimented in many fancy dives, never fearing the occasional back flop, etc.

According to a letter to my parents I had finally mastered the back dive, front and back flips, jack knife, swan dive, flying-dutchman, and half twists. This provided many hours of fun, and competition. We later built a 10 foot high platform with diving board.

(Clay Tries a Swan Dive)

We also managed to obtain a 4-man rubber raft with which we could paddle in the lagoon, and from which we could even fish. One day we discovered a large school of fish at about 15 feet depth, 50 yards off shore. As we would paddle to it, the school would dart off to another location, so we would follow it, with the school returning to the previous location each time. I decided to find why these fish were acting like this, since we were not doing anything to scare them. Instructing the other guys to paddle toward the school, I dove to the sandy bottom, and waited there. As the raft neared the school, sure enough they darted in my direction, as though fleeing something.

Suddenly, there appeared a large form swimming swiftly right toward me. There, zooming within arms length of me was a beautiful seven foot blue tuna! He sped past me in hot pursuit of the school of fish, scooping them up in his open mouth. I was enthralled by its slick beauty! No wonder the fish were fleeing this monster; they were swimming for their lives.

The tuna didn't seem to be annoyed by my presence, and did not appear to be at all dangerous. I felt right at home with this beautiful creature of God's lovely creation. Diving in these waters was providing me experience with many water denizens.

(My Underwater Adventures Below the Raft)

Our swimming adventures took many twists and turns, as we thought of other things to try. I had developed the ability to remain under water as long as 3 1/2 minutes without breathing. I had perfected a method of forcing oxygen into my lungs in saturated amounts by deep breathing and then compressing my chest muscles to force oxygen into my lungs.

I would take about three deep breaths like that, and then dive under water. I did not know at the time that it is possible to pass out while under water, by compressing too much oxygen into your blood stream. Its a dangerous practice. So one day while exploring the bottom at 30 feet depth, I managed to stay there a little too long with a feeling of euphoria. Its a long, slow way to the surface, even swimming upward when you're out of breath. I felt desperate for air! I was desperately tempted to breathe in, even knowing all I would get would be water. As I plunged through the surface, my lungs almost burst from my sudden gulp of air! Oh, what a wonderful thing air is, when you need it so badly. I have since read how many experienced divers have drowned by the very thing I had just experienced. Its like an exhilarating oxygen drunkenness that overcomes the common sense of needing to breathe again, and they stay underwater too long, and just pass out, and drown.

That terrifying experience led me to discover another way to provide air to a diver at depth, so he could stay down longer to explore. Using a gas mask, I hooked up a fifty foot hose to it, with the other end fastened to a regular tire pump. Finding a volunteer to operate the tire pump I descended into the depths of the lagoon. My pumping partner was working hard, and every time he pumped, the mask would blow up with lots of air. I had to be ready to take a gasp of air at the right moment, because the mask would deflate immediately, due to the pressure at that depth. This worked O.K. for a couple of minutes, until the pumper got exhausted. Then he just stopped pumping! I waited to catch the breath as he would provide the next pump, but no further air came. I could not know, while on the bottom, what they were doing on the surface, but I finally decided it was no place to be without air, so I swam to the surface. You can imagine my lack of composure when I faced my pumper. He could have drowned me! Communication was much needed. Well, so much for that experiment!

Being a camera bug, I had a great urge to take pictures of the beautiful under water coral gardens in the lagoon. As a Seabee, the "Can Do" spirit prevailed. No such underwater photography equipment was then available to us, so I realized I would have to design a waterproof box, capable of enabling me to focus the camera, and to roll the film. I made a stainless steel box, with a watertight removable cover, and a glass window, with knobs attached to the adjustment parts of the camera. Every moveable part had to be water sealed. What a complex job, but soon it was ready for a dunk test.

I filled a 50 gallon drum with water, and put the box at the bottom, without any leaks, after several tries. Then I tried it with my camera inside. What a risk by trial and error. It was finally ready for the big test at about 15 feet, under salt water. I managed to take three or four pictures, and then realized I did not know at what footage to focus it. Then I noticed some salt water sloshing around my camera. Oh, Oh!

Back on land I saw the sad results, as the box had blown a seal, under the great pressure. My camera shutter was ruined by salt water. Thus ended my attempt to take underwater pictures. The film I did manage to take was completely out of focus, so did not turn out. My camera was sent back to the States, and my patient father had the shutter replaced, and camera cleaned of salt water, and returned it to me in like new condition in about a month. What would we servicemen do without the support back home?

(An Artist's rendering of my Underwater Adventures, recorded in our Battalion History Book)

I returned to my observation of coral gardens while floating on the surface, with my head underwater peering through my water goggles, or diving to the depths. On my time off duty, I would sometimes float a mile or so out into the lagoon, and stay there for as long as two hours. I was simply captivated by the beautiful underwater spectacle.

The fleet that brought the hospital ship to our shores with the first of the nurses, also brought the Marine outfit that had invaded our island. Curious to see what we had done to their isle, some of them came ashore, and we talked with them. They were completely amazed at what had been accomplished in the few months since they had rid this place of the Japs. They found it hard to realize this was even the same island. They now were on their way to invade Saipan. One of our shop machinists had a cousin in this outfit, and they visited each other. That Marine had been on five separate island invasions, and expressed his feelings that Saipan was to be his last, and would be his burial ground. Sadly, that is how it actually turned out. Our machinist mate was later able to visit the gravesite of his cousin on Saipan. Such is the personally devastating result of war!

The incident of a "Red Alert" brought reality to the gun crew in the battery which was a few yards from our shop. We overheard them talking about getting some gunnery practice by shooting at a barge that had been promised them as a target. We saw the large barge towed into position and anchored about 300 yards offshore in the lagoon. My eyes brightened when I recognized the type of barge it was. Down in the engine room would be a large cylinder of Plexiglas, just like I had seen in the supply depot, and which could be used to make some marvelous souvenirs. If the barge was only to be used for a shooting gallery, they wouldn't miss that piece of useless plastic.

So went my thinking. I would wait until I got off duty, in the evening, about sunset when it was getting dark, and I could be assured they wouldn't be shooting. Then I could just slip into the barge, without being watched, and could relieve them of that useless chunk of worn-out plastic!

(Barge Being Readied to Tow into Target Position)

Its peculiar how twisted our minds can get when we really want something that belongs to another. We can rationalize to deceive ourselves into thinking it will be all right. The consequences can be serious, but in this case I really did not think such action would harm anything, or anyone. Little did I know what would develop!

When off duty, and evening approached, I put my tools in my little bag which was usually used for collecting seashells, and slipped into the waters of the lagoon, a short way down from the shop. I swam out into the deeper waters as I would usually do on my coral garden dives, and approached the barge from the outer direction. I could not see the gun emplacement, so assumed I was hidden from their view. It was getting dark rather fast, and I must hurry so I would have enough light to see by when down in the hold of the engine room.

I climbed over the gunnel, and slipped onto the deck and all was well, so far. I crawled down to the engine room hatch, and entered. It was getting dark in there by now, but I found my way to the engine, and just as I had expected, there was the Plexiglas cylinder! It measured about 18 inches long by 6 inches in diameter. All was ready for the capture, but it would require some unfastening.

Being unfamiliar with the engine, and its complex mechanical design, I didn't know what function this cylinder played, except I could see it was filled with water. Out came my tools, and I managed to get the screws and bolts loose, and removed, but the thing still seemed to be stuck on something, or being held somehow. I tugged and pulled with all my might but it still wouldn't budge. I could then see it had gasket cement that was holding it. A few jabs with my screwdriver, and it loosened. I figured it would now just slide endwise off its mountings, so I gave it a fierce tug, and sure enough the entire cylinder came sliding off clean as a whistle. Wow! What a surprise!

A geyser of water ejected with tremendous force, straight up to the ceiling. And it continued! I had opened a large hole directly through the bottom of the boat! Something I hadn't counted on, was that the cylinder had been fed seawater directly from the outside water, and when I released the cylinder, the full pressure of the water was then released. It provided the water-spout with pressure equal to the weight of the barge.

What could I do now? I immediately tried to shove the cylinder back into place, but the water pressure would not permit it. While I struggled, all I accomplished was to squirt myself with that geyser of water; in my face, body, and all walls and ceiling of the engine room.

I remember thinking at the time, what a silly thing I had done, just for a hunk of plastic. Now I had myself in a terrible predicament! In desperation, I just decided to forget it all, and depart the scene as expeditiously as possible. It wouldn't do any good to hang around there any longer. Gathering my tools, I fled the scene, with water gushing into the barge. Leaving the precious cylinder where it was, I slipped over the backside of the barge. Diving deep, I swam underwater as far away from the scene as possible to hold my breath, thinking all the time what a stupid thing I had done, and it was all for naught!

I was too ashamed and frightened thinking of consequences, to tell anyone about my episode. I was bothered in my sleep all that night, and could hardly wait to get to the shop, and view out on the lagoon what had happened to the barge. Without saying anything, my heart almost stopped when I saw there was no barge in sight! Then I heard a loud, excited shout go up from the gun emplacement Marines: "Hey, Sarge! Where's the barge?"

Then I was very much aware the worst had happened, the barge must have sunk to the bottom. Imagine, one of our men sinking one of our own country's ships! You have no doubt heard of "friendly fire"? It was a dreadful thought, and greatly fraught with dire consequences. Probably the worst that could happen was death by Marine firing squad. But more likely a court martial, with dishonorable discharge. The very least could be a captain's mast, with loss of rank. But, on the other hand the most likely punishment would be a real reading-out by our commander, (I rationalized) because after all, the Marines were about to sink it anyway by target practice.

The only way to actually prove the barge had sunk, rather than just floating away on the tide, would be to see it on the bottom. So when work was done for the day, I went into the drink for my usual swim, and out to the region of the barge. There in about 30 feet of water was the lost barge, sitting just as pretty as a picture, resting on the sandy bottom! One less barge the Marines would have to sink when the war ended.

It was decided by me, myself, and I, in a secret consultation, not to disclose any part of the story to anyone. But as a well known commentator would say: "Now,—here's the rest of the story." There you have it.

> In the scriptures, 1 John 1:9 says: "If we confess our sins, He is faithful and just to forgive us our sin and to cleanse us from all unrighteousness; and the blood of Jesus God's son, purifies us from every sin." I'm so glad I believe that, and have acted upon His great forgiveness.
>
> —A VET

(Captain Cruise Confers with Admiral Nimitz, on Eniwetok)

(Sunken Barge Site)

Marine gun emplacement was just to the right of this snapshot.
Our laundry and bathroom facilities show in the foreground.
My foxhole was under the pipes, at right.

MAROONED!
Alone on a Tropical Isle
EXCEPT FOR JAPS

Chapter 6

Marooned; Alone on a Tropical Isle (Except for Japs)

"Lessons of life are sometimes learned under harsh circumstances. We often follow the path of our past, careless practices, when God is showing us a new and better way. His!—**His way is perfect!**"

—A VET

Following our invasion of Eniwetok, our 110th Navy Seabee Battalion's first assignment was to construct an airfield that could handle up to the largest 4-engine bombers. In an amazing fete of engineering, and committed construction workers, we had a strip that allowed single-engine fighters to land, in only three weeks following our invasion.

(First Plane Lands on our New Airstrip)

Then one week later they flew in twin-engine "Billy Mitchell," B-25's, and just one week later still, they landed 4-engine, B-24's. The area had been cleared of explosives, dirt and trees, graded, crushed coral hauled into place, and leveled, and rolled down into a strip smooth as cement. They then watered it with salt water to keep the coral alive and growing into a smooth-packed solid strip of coral concrete, 6500 feet long.

This all required dedication and hard labor from every type of trade represented in our battalion. We all worked seven days a week, twelve hours a day, without a single day off for seven weeks. We had real incentive! Only 360 miles to the West, the Japanese were constructing an airstrip on Ponape, in order to launch planes to bomb us out before we could get our strip built. Even so, my spirit for exploration could not be dampened. I spent every available hour looking into new adventures.

The various facilities for our camp included several innovations that had originated in Seabee outfits that preceded us into the Pacific theater. Our drinking water came from a desalination plant set up by our Seabees, converting salt water into fresh distilled water, which was then stored in "Lister Bags." These bags were made of canvas, with a spout at the bottom. The canvas would "sweat," and evaporation of the moisture clinging to the bag would cool the water in the bag to a temperature almost like from a refrigerator. The water for showers came from shallow wells drilled into the coral. This water was partially salty because of seepage from the sea; not fit for drinking, but fresh enough for body cleansing, when using salt-water soap. It rained almost every morning, so for shaving, and washing our hair, and face, we collected rain water in barrels, and did our washing from our steel helmets. These were improvements indicating we had progressed greatly from our early days of pioneer island invaders.

(Our Washing Machine)

We constructed wind-driven washing machines by rigging large paddles, or 4-bladed propellers from cutout oil drum tops and bottoms. To this we attached a shaft made of pipe, with a cam section to which was connected an arm with a plunger at the bottom. This reached down into a large G. I. can (trash can) which would hold the wash water and our clothes. This was all built into a large wooden framework. The trade winds blew constantly, so we had a ready-made source of power. This rigging was a Seabee innovation, and I had the privilege of designing and building the one for our Machine Shop crew. Chief was already assigning me the special innovative projects, because none of the others cared to envision the unfamiliar design.

The Marines were not quite so innovative, and did not have the equipment available, so they washed their clothes in a more primitive manner. Tying their clothes to a rope, they would fasten the rope to a rock on the edge of the water, and expect the gentle waves lapping on the beach to eventually rinse their clothes, and cleanse them by salt water. It was quite effective, except when the waves got too big.

(Salvaged Clothes)

Then their clothes broke loose and disappeared into the lagoon, sinking to the bottom sand, 15 feet under water. Lost!

They were lost to the owners until I came along with my under-water goggles, and started salvage work, discovering their lost clothes, often covered in the sand, days later. With no way to determine the owner, I would take them back to our tent and distribute the clothes, depending on sizes, to whomever could fit them. We were well supplied with Marine fatigues for some time.

(Our Surfing Reef at Low Tide)

During our off hours we often swam on the ocean side reef, at high tide. The reef was about 100 yards wide, and at the edge of the reef the waves would break and would be white-water, rushing surf, all the way to the beach. We enjoyed surfing on these waves, but the coral was sharp as glass, and sometimes we would scrape our knees on it, with gashing wounds resulting. So with a carpenter in our tent, we designed a large surf-board with canvas covering.

Once built, we used this to ride the surf, usually saving us from coral cuts. This painful experience taught me the method for healing coral cuts. The coral contaminates the wound with a poisonous substance, but salt water will burn away this contamination. Other guys would go to our battalion doctor who would give them a salve to put on the wound, bandage it, and order them not to go into the salt water until healed. This was not for me, because no one was going to keep me from swimming if I could do anything about it! Meanwhile I would continue to swim, and in two weeks my wounds would be completely healed. The guys who had the doctor's prescription still had open wounds after a month. Salt water is really a good cleansing agent if you can stand the pain of its initial application.

My water goggles were the only ones in our battalion except for one facemask that belonged to our "Recreation" department. I put mine to use during most leisure hours, swimming in the lagoon. However, I was anxious to do some new exploration.

Possibly there was another island I could explore, and discover a new source of seashells, or coral gardens. It was at this time I heard of an island about 20 miles up the lagoon that was used as a daytime shore leave for fleet personnel. It sounded interesting enough, and quite secure, besides being some place else to visit. Our island was getting rather cramped. This new island was named "Runit," or Hawthorne Island. When we finally were given a day off work, I thought this would be a good challenge, and found that a taxi-boat stopped by that island on its way to the Northern-most island of Engebi.

Not needing any further urging, I packed a lunch, and *without notifying* any one I boarded the taxi early in the morning. It was a beautiful day, and I was equipped with my water goggles, swimsuit and sun helmet, besides my clothes, and shoes which I would wear on the reef. Twenty miles is quite a distance when traveling in a landing craft, but as we passed numerous smaller islands looking like I had imagined a tropical paradise to look, my excitement built.

Then a chilling thought crossed my mind. How could I have forgotten? I was now A.W.O.L. I hadn't thought to get permission from my Company officer to have a leave to exit our island! I hadn't pulled this stunt since leaving Hawaii! Then, it was a deliberate ploy to see Waikiki. This time it was an inadvertent oversight! I wondered if the authorities would look at it that way? It's too late now, so I might as well make the most of it.

Pulling onto the beach, they let me off, to explore on my own. There was a few Navy personnel who had already arrived from off a ship. They were preparing a picnic and games for other shipmates, to arrive later. I decided to find out a little of what the island offered.

The island was about a quarter mile wide, and one and a half miles long. The Southern two-thirds of it was thick jungle of palm, and breadfruit trees, and thick brush, with a very narrow strip of beach on both the lagoon, and ocean side. The place where I landed had a wide clearing, where two Quonset huts were under construction, intended to house ship beach parties, and indoor recreation. The remaining Northern portion of the island was what looked like a coconut grove, neatly tended, compared to the looks of the jungle area.

I was warned by an officer not to stray into the Southern jungle unless I wanted to risk my life by possibly running into a Jap still in hiding. I could see it would be almost impossible to see anyone even a few feet away in the jungle. This island had not been bombarded, and only a token landing party of Marines had set foot on it during the invasion, declaring it "secured." I figured that surely no Jap was going to expose his hideout to a large party of reveling sailors. So no one risked entering this one-mile strip of jungle, for it was "off-limits."

The shore parties were there for only the day-light hours, and evacuated the island at night. Since I was a Seabee, dressed in green fatigues and clod-hopper boots, I stood out like a sore thumb to these Navy blue-jackets. We didn't get along very well with these "regular" Navy guys anyway. So I went my own way, and they went about their antics. They all stayed on the lagoon side, so the ocean side was empty of personnel. That's where I chose to spend my time. Clothed in only my swimming trunks, a knife in holster, and my boots, I ventured out on the reef, with water goggles for viewing under water.

(Runit Shell Reef to be Explored)

A mile and a half of reef requires quite awhile to explore, so time seemed to pass quickly. I took some time out for lunch, and then back into the water for more reef-combing. I found a great variety of shells, and the fish seemed oblivious of an intruder. I would put the shells into a canvas bag hanging on my holster belt. Working my way back North along the reef, I was almost back to my starting point as I emerged from the water. I heard the revving of a diesel engine and recognized it was my taxi-boat pulling out on the return trip back to Eniwetok Island. I jumped out of the water, grabbed my clothes, and ran as fast as I could toward the lagoon. But alas! I could hear the boat already well on its way, and I was still a quarter mile away. My heart sank! So low it was about to come out my shoes! Now what could I do?

Arriving back with the sailors, I dressed, and watched them folding up to leave. Each ship provided its own whale boat for transporting their landing parties, and these were starting to arrive from the ships. Someone announced to me that my taxi-boat had already left. As if I didn't know! I was toying in my mind what alternatives I had. It was too far to swim back. But maybe if I could get a ride to one of the ships I could swim from there. But it would be dark before I could swim the four or five miles from the ship. I felt desperate!

The crew from the "Tennessee," hearing of my plight, offered me a ride to their ship so I could spend the night aboard, and return next day. But I had a situation of being absent without leave, and was afraid the ship's officers would report me. And I was conspicuously not "regular" Navy, in dress. So I declined their kind offer, but they did give me a left-over sandwich to eat. When they shoved off I felt like I had lost my last friend. What a "LOST" feeling; future totally unknown and dreaded!

I also knew the taxi-boat from my island was not regularly scheduled, so I couldn't be sure when they would be back, but hoped it would be on the morrow. Boy; was I ever praying: for safety, for food, for transportation, for not being put on report, and especially for God to forgive my foolishness.

At times like this we rapidly and forcefully learn our lessons, but meanwhile we learn patience while we wait with dreaded anticipation of how it will all turn out. Ahead of me was a long, dark night, alone, with possibly Japs lurking in the jungle while they waited for all to leave so they could raid the trash cans for leftover, discarded food.

(Runit Isle, Site of Marooned Seabee)

When the last of the sailors had departed, I started my search for a "hole to crawl into" for the night. One of the Quonset huts was all enclosed and they were building the interior counters, and room dividers. Behind a counter at one end I found packages of insulation which I could use for bedding, and for hiding. In the center of the hut was a screen-door swinging in the breeze. I figured if anyone came in the door I could hear them walking across the creaky, wooden floor, and I was positioned as far from the door as possible. I had my trusty battle knife on my side, but hoped I would never have to use it. The thought of possibly being that close to a Jap so that a knife could be used gave me the chills just thinking of it.

Under any other circumstances, this evening would have been like a dream, not a nightmare. The moon was bright, shadows long, palms blowing gently in the tropical breeze, with a fresh smell in the air. Only about 12 hours from now it would be daylight again.

Oh, if only those long and dreaded hours were already past. It's strange how the imagination works. As I looked out the Quonset hut windows I could imagine moving shadows that looked like men approaching the beach from the jungle. Then I would swallow my heart again, when I realized it was only palm fronds waving in the moonlight. Tiring of that preoccupation I retired to my under-counter hideaway to pray some more.

I don't know how many times I entreated God with the same request. He must have tired at my lack of faith! I dozed off—only to be awakened suddenly when the screen door swung closed. I listened tensely! I was sure I heard bare feet creeping across the floor toward me. My heart was beating so hard I was sure an intruder could hear it! No more footsteps. The door slammed again. Did he go out? I wasn't about to move and look out!

More prayer. Then I dozed off again. This seemed to go on endlessly, with the screen-door waking me so often. I wasn't about to leave my hiding place to block the door, to keep it from swinging. Only a few more hours. Several times I was sure I heard someone quietly walking across the wooden floor. In the darkness I was quite sure they couldn't find me, but what suspense. Never had time gone so slowly! Then in exhaustion, or in God's tender mercy I fell soundly asleep. The dawn was breaking as I awoke. What relief! Just to see the dawning of another day, by God's grace! I didn't venture out until I heard the sound of an approaching whale boat. I greeted the arriving sailors as though I was a welcoming party. Then for breakfast I had coconuts, and for juice the coconut milk. I was too proud as a Seabee to ask the Navy sailors for a food handout! Silliness to the extreme.

My enthusiasm for skin diving and shell hunting took a back seat to my vigilance looking for the taxi-boat. It finally arrived at 4 o'clock in the afternoon of my second day on the island. I clambered aboard almost before it touched shore. Finally I was on my way home again. So this is what it feels like to be marooned, I thought. You can have it! It's not at all like the romance I felt when I read the books of Robinson Crusoe or the Swiss Family Robinson.

I was still facing the consequences of my being absent without leave. Possibly my tent-mates had reported me missing? Maybe the Lord would let me get away with it. What mercy He would be showing if He did that! Well, the moment of our docking back at our pier on Eniwetok arrived. I was so *exceedingly* happy to be back safely. I thanked the Lord for keeping me from harm and bringing me home again. Scrambling fast to my tent, I caught my buddies just in time before they reported to the authorities I was missing. They "rattled my cage" for not letting any one know my planned trip, but then all was forgiven and we went to evening chow together. I'm sure I enjoyed my meal far more than anyone else that night! Two days without our regular three meals a day was a drag.

From then on, it became a habit to take proper precautions when embarking on one of my adventurous excursions. At least that was the case *most* of the time.

God doesn't just 'look the other way' when we sin; He some times allows us to wallow awhile in our own folly to learn the lesson he is teaching us. He is a just God—full of mercy, and tender love!

Psalm 101: 1 "I will sing of your love and justice; to you, O Lord, I will sing praise."

—A VET

Eniwetok
Fantastic!
CORAL GARDENS

Chapter 7

Fantastic Eniwetok Coral Gardens

On *very* rare occasions in life we are offered only one glimpse of something so remarkable, so magnificent, so startling, that it appears to be a peep through the windows of Heaven. Too wonderful for more than a brief glimpse, so as not to distract us from the responsibilities of our present life. Then as suddenly as it appeared, it is gone, never to be seen again by mortal beings!

—A VET

(A tiny Coral Isle. One of a Row of Islets)
(Surrounded by Fantastic Underwater Coral Gardens)

Swimming activities took on increasingly greater priority as our duties on Eniwetok were changed to island maintenance, and our hours were cut to 10 hours per day, and six days per week. The oppressive 120-degree heat during the day, and a cooler 90 degrees at night, was telling on our metabolism, so we often sought relief from the heat by being in the water.

A welcome relief from the routine of shop work came when I was assigned to work on a special project of constructing a dry-dock ramp on our neighboring island Parry. It was a half hour boat ride each way, and it got me off our island, besides getting to work with some guys from other Seabee outfits, Coast Guard people, and even some Navy pilots.

The ramp was to be used to pull small watercraft out of the water and up onto dry land so as to make repairs on their under-water hulls. It was also to be used to pull amphibious aircraft out of the water where they would land. We had to build a wheeled carriage, which ran on rails from deep water, onto the shore, and up the beach slope to the repair station on level ground.

My first introduction to their machine shop was a pleasant surprise. After working in our little shop trailer, I was escorted into an extra large Quonset hut normally designed to be an airplane hanger, and at one end was the largest machinist lathe I had ever seen. The bed of the lathe must have been 12 feet long, with a chuck that could handle a 36-inch diameter piece of metal. This was to be my piece of equipment to work with for the next couple of weeks. I was glad that my six week course of advanced training at Camp Perry had given me at least some experience on working with strange machine shop equipment, and larger then usual Navy machines.

My first task was to turn out steel wheels for the carriage that would run on a rail track. Then followed special parts and pieces of the latch mechanisms, and repair tools. We were kept busy for most every day, but occasionally we had to wait while assemblies were completed. During one of these periods of waiting, the Chief in charge gave three of us a number of hours to explore the island. This island was not nearly so inhabited as our own, and the northern portion was completely undeveloped. So that was the place we decided to explore.

As Seabees we were always looking for souvenirs, so with my trusty battle knife in its holster we set out toward the interior, trusting there were no Japs left here. We investigated a dugout area, and seeing a can half buried in the sand, we started digging around and sure enough, we had stumbled upon a hidden cache of Japanese foodstuffs. With my knife we opened one of the cans from a case of them, and looking up at us was a pair of fish eyes floating on top. The smell wasn't very enticing either, so we decided not to see what it tasted like. We also found bottles of what we guessed was sake, a Japanese liquor. (Wouldn't we have been surprised if it was TNT!) We left them, and the other foodstuffs, after looking around for something we could keep for remembrance, but found nothing. Thinking back we should have kept a label, or such, but I believe the cans were stamped with printing directly on the cans.

Moving on, we continued inland, and then encountered something unexpected, and frightening. A yellow band of ribbon was stretched around the area we were leaving. On investigation as we stepped out of the marked-off area, we saw a warning sign which read: KEEP OUT, ACTIVE MINE FIELD! We had inadvertently come right through the minefield, and had not encountered any explosives. The Japanese had probably mined their food cache, which we had so carelessly investigated, and we had been very fortunate. You can be sure, I thanked the Lord again for keeping us safe in our folly! What would we do without our guardian angel watching over our every move? Well, our adventure soon ended for that day. But I had seen enough, so that I had my exploration spirit awakened and desired to see more of this island, but it would have to wait for another time. So back to work we went.

While we were occupied on this job, one day I saw a most majestic sight! One of the huge amphibious flying boats, a Martin transoceanic transport, China Clipper, landed on the lagoon and taxied up to our dock area. I had never seen such a flying giant.

(Passenger "Clipper" Lands at Eniwetok)

After the crew had refueled, and rested awhile, they taxied out for takeoff. We watched their takeoff run into the wind, and it must have taken them at least three miles before they broke loose of the water, and lifted into the air. It was certainly a magnificent sight to behold! Before the war I had read so much about these planes, and seen their pictures in "Life" magazine, and I had never dreamed of ever actually seeing one of these planes in action.

With the job completed on Parry Island, I was back on Eniwetok again. It was kind of a relief, because there had been a rumor that I might have been transferred to the other outfit. For they had liked my work and needed someone to work on the large machines, but I had no desire to leave my friends. We still had plenty of work on our own battalion's equipment.

(Parry Island Dry Docking Ramp, after Completion)

Again, my desire to explore Parry, led me to take the taxi over there on my day off, and I headed alone, to the beaches on the North end of the island. This is the area where we had seen the Japs running through the trees the morning we had entered the lagoon for our invasion. I was right opposite Lady Slipper island, looking across the deep passage entrance to the atoll. The tide was still going out, but the reef was bare, so that I could walk out, clear to the deep-water trench.

As I gazed over to the other island my mind started to work overtime. Such a beautiful tropical island should be explored. It was only one and a quarter miles across to it, and the enticement was almost too much to overcome. I could easily swim that far. I wrestled with the thought for awhile, and even started taking my shoes off for the swim. However, something down deep inside of me seemed to say: "Don't do it!"

That got me to thinking. This passage was the link between the open ocean and the lagoon. Suppose a shark was laying in wait somewhere out there. Inside the lagoon the dolphins kept the sharks out, but out there by the ocean a shark might even dare to venture. The tide was flowing toward the open ocean, so that would set up a current. I picked up some wood on the beach and walking out on the reef, threw the stick out into the current. It went frighteningly fast out to the open water, indicating at least a ten-mile per hour stream! Well, that settled it for me; I wouldn't venture over to that other island. At least not by swimming.

(Fuel Dump on Lady Slipper Island)

That didn't end my desire to see Lady Slipper island though. There must be another way to get there. So in preparation for my next day off work, I inquired and found some men who had duties over there. Arrangements were made to go with them on Sunday, my day off.

(American Cemetery of Eniwetok Casualties)

It was about an hour's ride by boat, and it was a beautiful day, so with a lunch and my camera, we landed on Japtan, renamed Lady Slipper island. This was untouched by Naval shelling, so the tropical growth was unmarred, except for a fuel dump area, and our American cemetery, where the men were buried who were killed during the invasion, and other lines of duty. Our fire Chief, Jim Stewart, was buried here following the ammo barge explosion.

(Grave of Our Fire Chief)

It was very moving emotionally to see the graves, with white crosses marking them, with their names painted on the crosspiece. These were young men cut down in the very bloom of life, never being able to enjoy manhood. I prayed over those graves, that our ever merciful God would bring this conflict to a quick, righteous conclusion, and to bring comfort to those relatives and sweethearts back home who would be grieving for their lost loved ones. It was made even more real as I remembered seeing some of the survivors returning from the front lines that first day as we landed on the beach-head, here on Eniwetok.

Even today, as I write this my heart is wrenched, thinking of those who failed to return from that war. I have been so fortunate to have been privileged to return and marry my sweetheart, and to raise a blessed family, and to even get to see my grand-children. God has been *so graciously good* to me!

During my wanderings, I gazed across the lagoon inlet to Parry Island, and saw the area from which a few days earlier I had contemplated swimming across to this island. It had been so much safer to come by boat as I did this day. God leads us if we look to Him, as prayer is our means of communication. The tropical beauty of this place was outstanding, and I drank it in while exploring. Such a short time ago this was an area of contested turf, as we wrested it from our enemy, but the ravages of war had spared this beauty.

(Looking Out Onto the Lagoon from Lady Slipper Isle)

To look out from under the swaying palm trees, across the white beach, onto the azure blue, glassy water of the lagoon, was something of which quiet dreams are made and the memory is indelibly etched in my mind. But these dreamy times must come to a reality end, and so I took the taxi boat again back to our own isle of dreams, and work.

Surprises seem to have the character of showing up at the most unexpected times, and as it happened, one day I spotted what looked like a heavy cruiser enter our lagoon, and anchor a couple of miles off our island. I knew my brother-in-law, Elton Whildin was on the U.S.S. Chester, so I wondered if by some grand coincidence this ship could be his. Arriving at the Port Authority Office, I inquired of the port officer if the ship sitting off shore was the Chester? He looked a little disturbed at me, and made it clear it was no business of mine to know where our wartime shipping was located. When I explained I only wanted to know because a relative was aboard, he softened, and indicated it certainly was the Chester. Well, this officer was also the authority to provide leave papers to allow me to visit Elton, on this ship. I immediately changed my clothes into Navy blue fatigues, and sailor cap, and prepared to take a taxi out to the Chester.

(Dressed-up to Visit Elton)

You can probably imagine how excited I was anticipating Elton's surprised look when I would show up on his ship, after not seeing each other for about a year and a half, back in the States. He had no idea where I was located out in this broad Pacific expanse, and would never expect me to show up on this Regular Navy steel deck of his! Wonders never cease!

The taxi boat pulled up to the ramp hanging from the side of the cruiser, and my composure began to lose ground, as I realized the protocol of Navy procedure must be carried out. I must salute the ensign (flag), as I came aboard, and also the Officer of the Deck. The ship was so large, and quite pointed at both ends, that I couldn't see which end was the bow, (where the flag was hanging). So I embarrassingly ended up saluting the stern of the ship, opposite the flag! The officer didn't seem to notice, as I saluted him, so that sort of set me at ease. When asked my business aboard, I told him I was looking for my brother Elton Whildin, he immediately called a sailor over, and introduced us. He explained, they were all preparing for a Captain's Inspection, so I would find my brother working, but all sailors were to be given the remainder of the day off duty, after the inspection.

The sailor took me below decks, to Elton's compartment. As we walked through the hatch into his room, the sailor called out: "Whildin!" Elton swung around with a gruff voice: "Whatta ya want?" He stood there in amazement as he looked at me! His mouth dropped open, and he shouted out: "Clayton!" It was laughable to see his astonishment.

He had been grumbling to himself at having to do this cleanup work in preparation for a useless Captain's Inspection! Didn't they have more important things to do to bring this war to a close. This was September, 1944, and they had just come from the bombardment of Saipan Island in the Mariana's, and had been just offshore from Tinian, where our battalion was about to go for our next adventure.

After greetings, and being relieved from his cleaning chores, he prepared for the inspection, and they all mustered on top deck. Meanwhile, I wandered to the Marine detachment, which was the security guard complement of this ship, and therefore did not partake in this Captain's display. As usual, we got along very well together, a Seabee, with a bunch of Marines.

After inspection, Elton was dismissed to join me again. So he was my escort to show me about the ship. He had access to the forecastle, which is several decks above the main deck and is where the top officers on duty usually hang out, but today they weren't there. We were able to use their powerful binoculars, to look at the island, and Elton was able to see where we lived, as I pointed it out. We also spied on the nurses on the hospital ship which was nearby! It was so very interesting to see the insides of this huge warship, the largest one I had yet boarded. He did a great job in explaining what was performed in the various areas. The freedom of access on the ship amazed me, because I knew how strict the Regular Navy was with the military rules. Elton explained it was informal only because they were off-duty, as result of the Captain's inspection.

We went to chow in their big dining compartment, to a *wonderful* meal, the likes of which I hadn't seen in many months. He said it was again only because of the Captain's inspection, and it's holiday spirit afterwards. For a second dessert, we visited their ice-cream parlor, and had delicious sundaes; ice-cream which we never got while on the island. It was like heaven on the Eniwetok lagoon! Elton couldn't believe my elation, as he kept saying it wasn't always like this, that they usually ate like we did. Oh, how wrong he was, because these were the days we were eating C-rations every day as our mainstay. But I really enjoyed this change! We bid a reluctant farewell, not knowing what the future held, and I was on my way again, back to my home territory. What a refreshment it had been to again be with a relative, who knew the same people back home. He was able to catch me up on what my sister was doing while we were out here gallivanting around the Pacific, as if we owned it!

We each had further duties to perform while we helped bring this stinking war to a just conclusion. He, a shipboard sailor, and myself an adventurous Seabee, basking in the sun, when not swimming. Yes we really did work, too.

Knowing that our time on this island would soon be ending, and not being aware of what lay ahead, it was natural to want to experience anything new this place had to offer. By this time, my underwater adventures had become quite well known by even the officers of our outfit. One day I was approached by a lieutenant, and was told of an excursion a group of officers were going to make on the following Sunday. Would I care to go along? If so, they would be glad to drop me off at a group of five small islands on the Western side of the lagoon, where I could explore the unmarred coral gardens at my leisure. I said I would be overjoyed to have the opportunity.

I wouldn't even need to pack a lunch, because their stewards would be preparing enough food for all of us, with sodas for refreshment. Boy, this sounded like a trip made just for me! Of course I agreed to go along, and was all prepared at dawn on Sunday.

We gathered at the dock; about ten officers, and four of us enlisted men. They would drop us four off at our destination island, and then proceed farther up the lagoon to do some fishing. They divided the food, and we shoved off, in the landing craft, cruising about 15 miles to a chain of very small islands, with reef and sandbars tying them together at low tide. On the way we passed another entrance to the lagoon where they knew the water was about 200 feet deep, and with the smooth surface and extremely clear water, we could even see the bottom. What a fantastic day this was going to be for skin diving, with the water so clear.

The tide was mostly out, so as we approached the islands we could see that the coral was too near the surface to allow the craft to go all the way to the beach. No blasting had taken place to clear a path, so they decided to let us swim to shore. So holding our lunch above water we managed to swim and float on our backs to where the bottom could be touched. We set up a temporary campsite on the beach under one of the few coconut palms. The boating party agreed to pick us up in late afternoon, so we were now on our own to occupy ourselves at our individual pleasures. The other guys wanted to explore and fish, and I was there especially to explore the virgin coral gardens. With swimming trunks, and trusty knife in it's holster, and my shell bag hanging from my waist, I waded into the lagoon. No white men had ever been to this place, as far as anyone knows, so we each felt like true explorers.

I had seen some of the coral gardens off our home island, so I had an idea of the scenery awaiting me, but this was an unknown area, and could hold some unexpected secrets. My adventuresome spirit was peeked to the fullest anticipation as I approached the area that seemed from the surface to have a lot more awaiting me underwater.

As I dove beneath the surface, I was astonished how clear the water was, for I could see quite a distance, even though the water goggles did limit it some, especially in the peripheral vision. The farthest depth proved to be about 15 feet, and the coral garden filled most of the distance between the bottom sand and the surface.

(Typical Scene Near the Bottom)

The brain coral rises from the bottom, and is surrounded by little fish pecking their food from the ribs. Embedded in that type coral are the clams, which get in while small and then grow to giant size. Most of these clams are a maximum of 12 inches at the opening, and not the massive giant clams found around the Fiji islands. The colors of the tissues along the open shell are royal blue, purple and reds, and they spill out around the shell in a most fantastic explosion of beauty. They attract fish to the opening by their color and blending into the coral, and trap the fish as they snap shut.

As I have tried to put a stick into their opening, they clamp it so tight you cannot pull it out. One reason for having a knife along at all times while diving, is in case you would get a finger, hand, or foot caught by one of these clams. The only way to get loose is to possibly pry it out, or worst case to cut off your appendage to escape. So you proceed with care around them! But that's no reason to keep from enjoying what goes on around this coral. As I swam further into this forest of coral trees and giant coral ferns, it took on the appearance of a fairyland of beauty. Fish were everywhere around the growth. Most of them were about three to five inches in length, and of fantastic design of colors. Oh, how I wished my talents included that of an artist, but possibly in the future I could learn, so I tried to remember these scenes to paint them later.

In a letter written to my parents, I tried to describe the scene now unfolding before me, quote: "Believe me, you never saw so many shapes and colors of fish in any aquarium. It is wonderful, and so fascinating to watch them swimming around in the coral. The variety doesn't stop with ten or twenty different species, but are more like the human race, everyone different, and they all swim around in schools together. There are mixtures of bright blue, yellow, orange, red, white, green, brown, black, silver and all are matched so perfectly. And there is one fish that looks almost like a checkerboard with silver and brown squares. It's all just indescribable!" The colors are brilliant; never faded at all.

And as fantastic as the fish are, so also are the various types of coral, both in shape, and color. None of this is organic plant life, but calcified material, that looks like it was frozen suddenly in place. The fish swimming around through the growth were not in any way afraid of me, but they were very curious. They would swim right up to my hands, and nibble on my fingers. As I came around one coral growth, a large Grouper fish was looking straight at me as though laying in wait for my arrival. I paused, and looked him right in the eyes, when suddenly he darted right for my nose. He stopped about a foot in front of me and just starred at me in a very dumb expression. I burst out laughing, but only bubbles came out! I guess we were both surprised.

Coming up for air, is a requirement of course, and that also provides a chance to rest. The salt water makes it very relaxing to swim, because a person is so buoyant. I would swim to a new location and then dive to find another scene. I felt like those kids in fairy tales must have felt when they were exploring the forest. Or a candy city, or floating across a beautiful landscape on a magic carpet. It was just simply, "out of this world"!

Swimming through the forest of coral shapes, amidst the delicate denizens of this fairyland is an experience that is certainly indescribable! There is just nothing on the dry land that looks or feels like that, and words fail.

(Typical Fan Coral)

I was so very spiritually moved by this experience, that as I slowly drifted along, I communed with my Creator, in prayer. "Lord, your creation here is beautiful beyond description, just as I imagine Heaven to be. While you are there preparing a place for me to live with you, would you possibly consider doing one little favor for me in my mansion? I would love to have this coral garden, or one just like it, as an aquarium in my mansion. Then I could show it to my beloved friends and family, so they too could experience this wonderful beauty you allowed me to glimpse here on earth. This is Your creation, just as Heaven is. Lord, please?" I was deeply moved, and strange as it may seem, I expected God to answer my prayer! The 'rest of the story' awaits at the end of this narrative. No peeking now!

At lunch time we all gathered in the shade of the palm trees to share our lunch. It consisted of: peanut-butter sandwiches, cheese, cold meats, wieners, jam, boiled eggs, plenty of bread, and fresh apples. We enjoyed this picnic lunch like it was a feast. Then we each went back to our activities. I did some beachcombing for shells, and then back into the coral gardens for some more exploring. The underwater scenery was continually changing, with unequaled variety. I don't think I would ever tire of this way of exploring. But the day was drawing to a close, and we were expecting our boat to return at any time. This had been a most enjoyable expedition!

The officers arrived on schedule, and were proud to show us their catch of the day. It was drawing close to sunset as we pulled away from the islands. To our surprise, and excitement we were joined by a pod of Dolphins, swimming alongside the boat. They were cavorting off to the side, diving deep and then in formation flying through the air in an arcing jump, sliding back into the water without a ripple. The lagoon was as smooth as glass, and this was an extraordinary sight. It was similar to a SeaWorld act that would come years later.

(Sunset, the End of a Tropical Day)

As the sun set across the lagoon, the beauty was simply gorgeous! The rays of the sun glistened on the smooth water, and the sky took on a glowing red, that soon saturated even the water's surface. If only I was an artist, this would be a prize-winning picture! No color photography had yet been developed for our use.

We arrived back at Eniwetok Island after dark, tired, but so glad we had taken this outing. Indeed, this experience was a highlight of all my time overseas! And we had gained a better understanding that these officers were a bunch of regular guys, further cementing our togetherness as a Seabee unit.

Now comes the 'rest of the story': After the war was over, brought to a conclusive end by the dropping of two atomic bombs on Japan, our scientists continued to work on further development of atomic energy. Then the research was expanded to the Thermonuclear H-bomb! The Eniwetok Atoll was chosen as one of the sites from which tests would be made. So in 1952/53, a Hydrogen Bomb was placed on the little island on which I had eaten lunch, and explored the coral gardens. When it exploded, the very coral garden where I swam, and where I prayed, as I viewed it, was turned into so much energy: E=M (C squared). In words this means: Energy equals Mass times Speed of Light squared. The material of the coral gardens was changed into an equal amount of energy.

I can speculate, God needed to do that in order to transport it up to the New Jerusalem, to become an aquarium in my mansion, as he changed it back into a beautiful Coral Garden! An answered prayer? I guess I'll have to wait until we see our heavenly home to be sure of it. But my imagination runs wild at times!

Even miles away, the island on which I was marooned for two days, became so hot with the bomb blast that it melted the coral sand. It formed a cliff about ten feet high, along the beach. I saw the picture of it and it was incredible! In 1952, when I took my final exam in Quantum Physics, at the University of Wisconsin, we were asked to derive the equations to prove a *hydrogen bomb* was feasible. Then a year later the secret was out that the U.S. had succeeded in building such a bomb. Yes, I passed the exam! Later that same year, I became the Nuclear Radiation Specialist for Beckman Instruments. What goes around, comes around.

> It was indeed a privilege to view that coral garden scene, as probably the only human to have seen it, and then it was removed from the Earth environment, never again to be viewed by mortal man. God has even much more *actually* in store for us in Heaven than what I can imagination. For He said: *"Eye has not seen, nor ear heard, neither have entered into the heart of man the things which God has prepared for them that love Him. But God has revealed them unto us by His Spirit;—"* (1 Cor. 2:9)
>
> —A VET

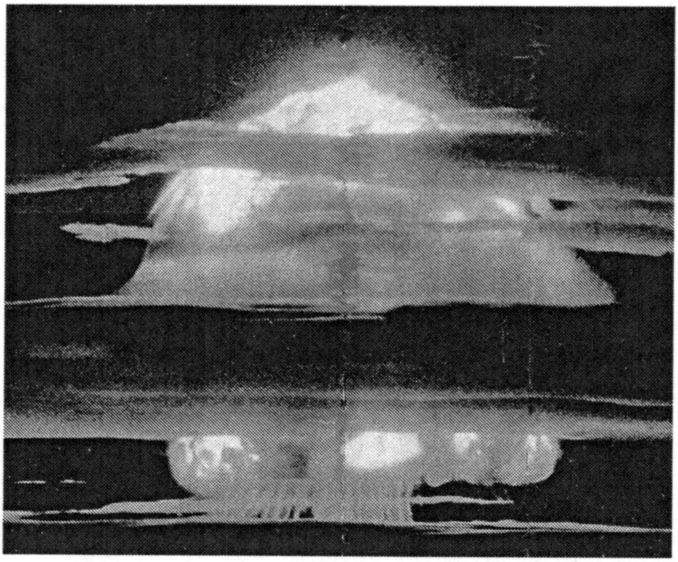

(The H-bomb Transmutes Matter into Energy by
Thermonuclear Transformation)

"Ace" in Reverse

Planes Rescued Under Fire

Chapter 8

"Ace" in Reverse or Planes Rescued Under Fire

An "Ace" is usually spoken of a fighter pilot who shoots down five or more enemy planes under combat conditions. We all may not have the bravado of those fighter pilots, but often in the heat of emergency, we perform feats we would be afraid of doing under any other conditions. Such was the tragic incident one dark night on Eniwetok, that censors wouldn't, under any circumstances, allow the news media to make known to the public. This account is now provided without censorship.

—A VET

It certainly is not unusual that on a three mile long island, with an airstrip taking up half that length and with hundreds of military flight operations daily, that many mishaps, involving aircraft and personnel would take place. We really didn't know what would happen as each day began.

Such was one incident as I was abruptly awakened at 7 A.M. one morning after working till two o'clock on the night shift. The sputtering engine of a Hellcat fighter plane, woke me as it seemed to be heading right for our tent. I jumped from my cot, and ran outside to see a plane faltering on take-off. Our tents were just to the side of the end of the airstrip, where the planes usually were well into the air at that point. That plane was only a couple hundred feet in the air, with its engine spitting and sputtering as I ran the hundred yards to the beach on the reef side of the island. I expected the plane to come down right about there, and wanted to be available in case the pilot could be rescued. Other men came, and we watched as the engine quit. The plane started to descend and we expected a crash, as those warplanes sink like a rock, without engine power. Then the engine coughed and started up again. He managed to regain a little altitude, as he raised his landing gear for a belly landing on the reef, which had about a foot of water on it.

His engine stopped again, and he lost altitude as a result. We were all trying to hold him in the air by sheer thought. But I was calling on the Lord to save that poor fellow. Our arms were even making the holding motion to keep him in the air. When only a few feet above the reef, the engine caught again, and he retained his level flight. We all cheered him on like a college crowd at a football game. At about twenty feet the engine quit. The crowd by this time that was gathered all along the beach, could only sigh with an, OH! The sympathy felt for the pilot was extremely compassionate, as we saw a man who presently was alive and fighting for his life. With a plane loaded with high octane aviation fuel we all expected a ball of flame to accompany the crash. But the engine again started, as he was almost touching the water. He gained a few feet, and with flaps extended, he cut the switch, stopping the engine and settled into the water while just above stall speed. A little splash and he slid along on the water cushion until he ground to a stop on the rough coral reef. No flames, as the water snuffed the sparks!

He immediately jumped from the plane, and splashed through the water on his dash to the beach, expecting an explosion. What a cheer went up from all the spectators! I knew that God had brought him safely down! They reported later that the plane was only slightly damaged, because of the pilot's expert handling of the emergency.

(Fighter Pilots Headquarters Area)

Later that same night, we shift our attention to a slightly different situation. We had several theater areas on the island, one for each branch of service. So for variety, I chose to attend the fighter pilots theater, which was showing the movie of Betty Grable, starring in "Pretty Baby." This area was in a secluded place across the airstrip from most other activity, located just under the Control Tower. It was a nasty night with pouring rain, but nothing could keep me from seeing my pinup girl on the screen. Besides, I had the efficient poncho, that Miss Cora Jacobson may have made back in the States as she worked in the Poncho production department of Rainfair, the main producer of this type rain gear.

The night was moonless, and with the rain falling heavily, driven by whistling winds, it was so dark you couldn't even see your hand in front of your face. The trick of getting to the theater area was to cross the airstrip. It was almost a mile to the end of the strip from where I was, and despite being against regulations, I decided to cross at the middle of the strip. Certainly there wouldn't be any flight operations under these weather conditions. So I thought! The strip was about 100 yards wide, and I was well out into the center, when suddenly a jet black form whizzed past my face. I realized it was the wing of a fighter plane, as he was in his landing roll-out. The wind almost knocked me for a loop, as I caught his propwash. I couldn't go back, as I had reached the point of no return in my crossing, so I took off running into the black hole ahead of me! Wow! What a narrow escape. I don't think I enjoyed the movie as much as anticipated, as my heart took some time to calm down. I thought, what more can happen to this innocent Seabee? If Miss Grable could only realize what we servicemen would go through to see her on the silver screen. Oh, well!

During the invasion of Eniwetok, in February, 1944, the Navy struck the Japanese Naval base at Truk, to knock out the airbase from which they could have launched an air attack on our invading forces. It was a simultaneous strike, to knock the Japs off balance. And even though great losses were inflicted, the base at Truk remained a threat to our forces. Consequently, after the planes flying from our island had bombed out our nearest threat at Ponape, 360 miles to the West, we then considered going after the Jap forces at Truk, which was 700 miles away. Our B-24's would go in at high level, because of the heavy defenses, but by bombing from high altitude were unable to do much serious damage.

(Truk Lagoon)

Along came a daring young pilot by the name of Commander Miller! He had a crew that was very dedicated to him, while these initial bombardments took place. We were hearing much about him from other pilots who hung around our shop on off hours. It seems he was plotting a special mission, which appeared rather risky from his peer's viewpoint. But contrary to military policy, the high brass was not being consulted. One day, taking off with a full fuel load, bombs, and ammunition, he went alone on a "search" mission to discover any Jap strongholds.

In about three hours they approached what he and his navigator knew to be the island of Truk, with its large lagoon filled with ships. Flying at low level, just over the waves, he caught the Japs completely by surprise. He came roaring in over the island's communication center, loosing a couple bombs directly on their facility with devastating aim, machine-gunning as they passed by. Returning over the harbor, his bomber laid open the sides of ships, before the enemy knew what was happening. Another pass over the main island facilities, did more destruction. By this time flak was flying all around them, as they made another pass over the ships while dropping the last of their bombs. It was time to get out of there, and that they did at wave-top level.

Amazingly, they had gotten away practically unscathed. But what destruction! I don't know the military flak he received from his Commander back at our base, but the result was a total change in bombing strategy from then on, in the attacks against the Truk stronghold. Later photos showed that his plane in this one attack had reaped more havoc on that base than all the previous high altitude bombing.

This was only the beginning of Miller's episodes. Attack after attack was carried out at low level, which was proving to be very effective. His crew, some of whom would stop by our shop, was getting very jittery, and some had been wounded, requiring replacements, which were not volunteers. Who would want to be a part of these suicide missions?

This reputation led to a nick-name being assigned; "Commander Miller, the One-Man Air Force"! One afternoon, Smitty, Gansky, and I were walking along the reef near the end of the airstrip, when one of these missions was coming to an end. Of course, we were unaware of the drama being played at this very moment in a bomber approaching for landing at the far end of this airstrip. It was Miller, coming in from another raid over Truk. His plane was all shot up, with no communications, and all hydraulics destroyed, flaps damaged, and with wounded men aboard. They had barely managed to limp back the 700 miles from their attack, and now were about to make a landing on our strip. This was to be a landing without any brakes, no usable flaps, and to stop the plane, he had instructed the crew to deploy the chutes out the side hatches as brakes. He would give the signal when to pull the chutes. However, as they touched down the intercom system went out, so that he could not communicate with the crew. As a result, the chutes were not deployed, and the plane continued down the strip, with little loss of speed.

We three Seabees saw the plane coming, and didn't know if it was taking off, or landing, but knew it was not going to stop by the time it reached the end of the strip. We watched in horror as it jumped off the shallow ledge, plowed across the sand, collapsing the nose-gear, and skidding on its nose out onto the reef. As the plane came to a stop, the front hatch sprung open, and the flight officers came bursting out, expecting a fire. We ran down to help, but by this time the rest of the crew came out their hatches, jumping to the reef, and running away. No fire resulted! So a rescue crew arrived, and evacuated the wounded and took the rest of the crew away for a debriefing.

We later received a thorough briefing from other air crews at the favorite gathering spot, our shop. (An interesting sidelight came to my attention shortly after the war. It was by an article in the San Francisco paper, telling of the death of Commander Miller at his home after suffering from deadly consumption for the last years of the war. It became my opinion he was trying to die a hero, in his daring aerial exploits, because he had known his health condition at the very time he was performing such suicide missions.) It was a privilege to have known this brave patriot, and to witness his heroic action, on behalf of our nation which he loved.

An exceptional precedent was being played out affecting the life of one of my Seabee friends. This young man in civilian life had already obtained his Private Pilots Certificate, and was a capable light plane pilot. His prowess as a private pilot had come to the attention of our Commander. And even though only an enlisted man, as a 3rd Class Petty Officer he was assigned the pleasant duties of Liaison Pilot for our officers, flying the military Aeronca light plane while on Eniwetok.

(A Military Liaison Aeronca L-16)

He further came to the attention of the Navy pilots, and with exceptional pull, was offered an unprecedented opportunity of training to be a Navy pilot, while based there on the island. Our forces were in great need of pilots, and especially if they could be put into immediate service.

He had already advanced to the stage of flying solo in the Hellcats based on the island. One day while taxiing for takeoff, a fuel truck suddenly pulled in front of him, so he hit the brakes, and did a nose-over. The nose of the plane, and propeller were damaged, and so was his reputation! As a result he was ordered to a military tribunal called a "Captains Mast." As they investigated the incident, and his illegal training to become a Navy pilot, they began questioning about another unsolved mystery.

It had happened that a B-24 Bomber had made an unauthorized flight, and when landing, and taxiing, was told by the tower controllers to report immediately to the authorities. Two men were seen exiting the plane, but no one had reported to explain the unauthorized flight, even though the Military Police had been notified to apprehend the men. As it was disclosed in the hearing, this private pilot friend, and a buddy had taken the 4-engine bomber for a flight over Kwajalein island which is about 360 miles away, and returned, without being caught. Now the truth was out! He was washed out of the pilot training plan, busted in rank, and fined two hundred dollars. (That was 3 1/2 month's pay) He felt he got off rather easy!

The strangest adventures seem to happen to Seabees; could it be they are a special breed of humanity? Or did their training give them a special oomph? They were truly military men in the strictest sense but acted more like a civilian performing a military task; so they were different, in an unusual way! Personally I loved the challenge of adventure, and still do!

Another type of plane used for many of our bombing tasks was the B-25, Billy Mitchell bomber; a fast, twin engine member of our fighting arsenal. One of the pilots became a friend of mine, as he would come to the shop, having need for aircraft parts to be machined. As he wanted to thank me in some special way for a job well done, he offered to take me up for a mission hop with his crew. Prior to my flight he planned to have a new engine installed, and was taking the bomber on a mission about 350 miles away, to give it a good breaking in before he would include me.

(B-25 Billy Mitchell Bomber)

By coincidence, I happened to be near the airstrip when a B-25 was approaching for landing, having difficulty with lowering one wheel of his main gear. He was shot up, and only one engine was functioning. I watched his approach, and wondered how he would land. He touched down, with the one main wheel touching and the other wing held high. Gradually braking with the one wheel he managed to slow the craft, and keep it in a straight line, and then on losing speed had to let the other wing lower, and the nose wheel touch. As the wing contacted the runway, the plane went into a cartwheel, and spun around in a complete circle, sparks flying everywhere. As the forward speed slowed the plane stopped spinning, and came to rest near the side of the runway. The crew burst from hatches, and ran from the plane. No one was injured, and no fire erupted. Crash trucks soon reached the scene, and all was in control, preparing to move the wreckage from the active strip.

It turned out the pilot of this plane was the one who had promised me a flight with him. He came to the shop to apologize to me that he did not dare take me up until he had obtained further practice in flying with only one engine. He said he had flown 350 miles on only one engine after the other was shot out by enemy action, and had trouble keeping it at altitude because of lack of experience with this type of emergency procedure. I never collected on this promise, because our outfit moved out shortly after this episode. I'm glad I wasn't along on that mission. God spared me the fright.

The Corsair fighters were our favorites on this island. They were fast, beautiful in flight, and were flown by both Navy, and Marine pilots. As we were relaxing in our tent one early evening, we heard the familiar sound of a Corsair taking off, but as it neared our tent, the engine started to sputter, and bark. We knew he was in trouble, so running outside, and down to the beach we saw the plane several hundred feet in the air. The pilot probably figured he had enough speed and altitude, so he banked the plane around in a tight circle and headed downwind back to the strip. There was quite a strong cross-wind, and we also noticed flames and smoke coming from his engine, and enveloping the cockpit.

As we watched the plane losing altitude, we saw the pilot climb out of the cockpit, and get his feet on the wing, as he was hanging onto the joystick for controlling. But without rudder pedal control the wind was blowing the plane toward one side of the strip. In horror, we watched the desperate maneuvers of the brave pilot, as he endeavored to keep the plane heading down the strip.

Why didn't he pull his chute? Last moment attempts failed, and the Corsair crashed with the pilot, right on top of four B-24 bombers, in a huge ball of flame. The cross-wind had blown him right over the plane parking area. I'll never forget that scene, of a man fighting for his life, and then to see the tragic end!

(Corsair Crashes on Top of Four B-24's)

It isn't always the action of the enemy that causes death, and yet the enemy was the ultimate cause because he had started this tragic war. Here was another young man that would not return home to his loved ones. It really makes a fellow think seriously! Possibly that is why our private Bible study and prayer group numbers continued to grow. The Lord was drawing men to Himself!

The aviation arms of the military were one of the most dangerous duties, yet one of the most effective against the enemy. The danger was not only present by enemy action, but the danger was often imminent by events on home territory.

In the early morning hours, at about 2 A.M. on August 10, 1944, three of us were manning the night crew at the shop. We heard the familiar drone of four engines working at full power as a B-24 was taking off for a bombing mission, racing down the runway toward us. We could tell when it left the ground by the change of sound, so it was well into its climb, when we heard its engines cough. The sound suddenly changed to what we detected was the sound of only two engines. The sound became a strange whine, and then a sickening crumbling of metal, continuing with crunching, banging, and scraping sounds. As we looked up we saw a huge ball of flame, and a terrific explosion accompanying it.

We immediately knew the bomber had crashed! It was only about 300 yards away, so we took off running toward it, to be of rescue help if possible. Being the fastest runner, I was approaching the flaming mass, on the far side of the parachute loft, when I was met by men running toward me. They were yelling: "Run for your life, the plane's loaded with bombs and they're about to go off!" I reversed my run as fast as possible, heading for the foxhole by my tent.

(My Foxhole Dugout)

I had run only a few yards when the first bomb went off, flinging me to the ground by the concussion. Jumping to my feet, I almost flew over the ground, arriving at my tent just as the second bomb went off, knocking me flat right in front of the screen door. My tent mates burst through the door just then, and fell stumbling over my prostrate form, like bowling pins.

I knew by that time they were all awake, so I ran around to the back of the tent where Trigg and I had our neat dugout, with boards and sandbags over the top of the foxhole. He shortly joined me, and we sat there while four more bombs went off, each one knocking more sand on us as the blast shook the ground. We were only 225 measured yards from the crash site, and it was really scary, not knowing but what an explosion might throw a bomb down on us. The blasts seemed to be finished as we waited, but very soon the loud speaker near our tent, buzzed and an excited voice appealed for any volunteers to approach the crash site for rescue work. The bomber had crashed right in the middle of the fighter parking area, containing all the spare planes for transfer to aircraft carriers at some later time. "We need your help, desperately!" They repeated.

Still not knowing how we could help, I pictured they wanted people to go into the flames to rescue flight crew, or any others who might have been involved on the ground. However, putting aside my reluctance, I climbed out of our foxhole, and ran toward the site of the blazing wreckage.

As I arrived, I saw our fire department crew spraying water on the bomber remains, and they asked for a volunteer to approach the unexploded bomb just outside the flames, to spray water on it to keep it cool. That was not for me, but one of our men, Lugo, volunteered and pulled a spray hose ahead of him to quench the heat. As he neared the bomb, he gradually turned the spray on it, until he had it all covered with spray. The firemen sprayed water on Lugo, so he could bear the terrific heat. It was remarkable to see his bravery, as he was only feet from the main fire. And he not knowing whether other bombs would go off, blowing him sky high, let alone the one he was spraying, which could go off at any time!

We could see the fire was burning all along the lines of fighter planes, gaining momentum by the second. Fire crews let us know all the fighter planes would burn or blow up as their fuel tanks were starting to burn. Each plane was also loaded with belts of ammunition, which could start exploding at any moment, from the heat and flame. The idea was for volunteers to rescue as many of the planes as could be reached, which had still not caught fire, and to push them out of the fire path.

So in the face of advancing explosions, we raced to any plane we could approach in the heat. Five or six of us would get on one plane, and maneuver it around to where we could push it to a vacant area. These were all carrier aircraft, and had their wings folded up. Having gotten one plane to a safe haven, we would select another to rescue. We worked with the same volunteers as a team, and together figured how best to get the planes away from the flames.

As we were pushing the fifth plane past the blazing bomber, suddenly we heard more of the ammunition exploding, this time very near to us. As my back was toward the planes as we pushed, we saw 50 caliber tracer bullets whizzing past our heads, some striking our plane. We knew the tracers were every 5th one in the string, so we were being machine gunned with a hail of bullets. This was no place to be! So we took off running over to the beach embankment, and took refuge! We were satisfied that we had managed to rescue five planes from the flames, at least. Other volunteers had gotten some planes away also.

I figured my volunteer efforts were no longer demanded, so continued back to our tent. After discussing the event with my tent mates, I returned to the shop duty. It was impossible to do any more work that night, as men came by from every line of duty, each wanting to talk about their experience. We gathered more word about the happenings as the hours drew on. After the ammunition had begun exploding in random directions, not much service could be performed at the site, so the remainder of the burning planes were allowed to burn themselves out, as they were beyond repair.

(B-24 Crash Scene as it Appeared the Next Day)

The bomber, as it was taking off and had a couple hundred feet altitude, had lost both engines on one side of the plane, so it had gone out of control. It cart-wheeled down the entire line of parked fighter planes, destroying them or setting them on fire as it spilled fuel. In all, 130 fighter planes were destroyed or damaged badly, as well as the bomber. It had aboard a full load of fuel, full bomb load, and ten crew members. Three men had managed to crawl from the blazing wreckage, and had been rescued by witnesses, and delivered by ambulance to the emergency clinic. (In 1946, while attending Mt. San Antonio College, a classmate of mine turned out to have been a medic in the clinic where they brought the three crash victims, and he told me only one of those men survived. The other two had died of their severe injuries and burns.)

The next day we discussed and meditated on the event of the past night. After measuring the distance, and estimating the speed the bomber was traveling on takeoff, we ran calculations. If the plane had remained in the air only two seconds longer, and traveling in the same direction, they would have crashed right into our tent area and destroyed our entire Company who were housed in that vicinity. God watches over His children, even when events are completely beyond our own control. He controls all things, and beings! This was just another occasion where my Lord had me in His loving and shielding hand. I praise His Holy Name!

ACE

Five Planes Rescued Under Fire

C. F. Rasmussen

Awarded This Day August 10, 1944 for Voluntary Bravery Beyond the Call of Duty!

("Imaginary" Award Plaque)

As a pilot becomes an "Ace" by destroying five enemy planes, we had become an "Ace" in reverse, by *saving* five of our own planes.

The Psalmist reminds us: "God shall give His angels charge over thee, to keep thee in all thy ways." We need not fear, for He has a plan for our lives, and nothing will prevent Him from completing that plan. He watches over us even as He prepares us for His service, kept for His purpose! Even as the Psalmist was kept by God as He prepared him to become Israel's King.

God sometimes calls us unexpectedly into service which he had prepared beforehand for us to do. We must be prepared. (Ephesians 2:10)

—A VET

STRAFED BY ZERO
Look Out!
"Douse the Lights!
--A Zero's Diving!"

Chapter 9

Face-to-face with a Strafing Zero or "Someone Douse the Lights!"

Never having faced such a situation before, it's difficult to say how you would react when the lives of you, and a bunch of friends, face death at point-blank range. Do you act for the safety of all or do you turn and run to save your own life? It happened one night in a surprise attack on the Island of Tinian.

—A VET

What a day of rejoicing it was when we received the word it was time to move out! We would soon be on our way to the next line of offense against the Japanese. The voyage would be about 1200 miles to another formally Jap-held island of Tinian, in the Marianas group. It seemed we always were required to be the point men in the progress of our armed forces toward the enemy stronghold. But this time it would not be my turn to be part of the first waves invading forces. It would be such a pleasure to leave this tiny island of Eniwetok, which was only three miles long, by an average of a quarter mile wide, with its average daytime temperature of 120 degrees.

We imagined that no matter what this new island was like, it must be much better than where we had been for the past seven months. Yet we realized it was much closer to the Japanese homeland, and reachable by enemy bombers from several areas. The first echelon of our Seabee Battalion to leave consisted of 270 men, who were to help the Marine's mopping-up operations on the island of Saipan. They left on August 24, 1944, and arrived on Saipan on August 29. My friend from the shop, Hank Brunnetti, was in this group, and his cousin was one of the Marines killed in the invasion, and Hank was there to witness his burial. We had previously met this cousin on Eniwetok, and were all saddened to hear the news. This echelon, on September 9, moved to the island of Tinian, to take part in that invasion. On September 13, the second echelon of 290 men set sail, arriving in Tinian on September 18. The third echelon of 215 men departed Eniwetok on September 26, arriving in Tinian on October 1.

(Loading an LST—Departing Eniwetok)

I was to be in the fourth echelon, and by this time I was the only one left in our tent, so I felt quite lonely. We were assigned to load the LST on which we would be sailing with all the remaining bull-dozers, scrapers, cranes, jeeps, and trucks. This was now my opportunity to operate the bull-dozers! I would drive them up the ramp into the nose of the LST, and then an expert driver would take it from there. He would drive it at full speed down the sloping steel deck. As he neared the row of machines, he would lock one track, and spin the huge dozer 180 degrees as though on ice, backing into the slot where it would be tied down. I marveled at his skill in getting it to end up exactly where it belonged!

We were all now getting very anxious to be on our way to our new home. My best friend, Bill Dally, was with me on this leg of our travel. When the LST was finally loaded with all our equipment, both in the cavernous hull, and topside on the open deck, the huge front doors swung closed. We were then ready to depart at high tide, when they could back the now-loaded ship off the beach. Our Seabees numbered 270 men on this voyage, with only 25 men left on the island to clean up the last of our mess, before they also came.

Bill and I set up our bedding and gear under one truck on the top deck. This would keep us dry in case it rained, and would shield us from shrapnel in case we were bombed or strafed by Jap planes. We were traveling in enemy infested waters, with the constant threat of submarines as well as air attack! We were treated with unexpected friendliness by the ship crew, and the first chow call brought food like we hadn't eaten in a long time. The Mess-hall was on the top deck, entered by one hatch, getting our food in a serving line, then exiting from another hatch, all on the one deck. The food was delicious, and enjoyed by all. We even felt so thankful we went back to thank the mess-cooks. I drank a coke with each meal which I think acted like a tonic and to my relief didn't get seasick once!

We had heard of the dreaded typhoons, so wouldn't you guess it, we were struck by one our second day out. An LST is a flat-bottomed ship, and tends to bob like a cork on the waves! As we observed from the stern of the ship, when the prow was headed into the waves, we experienced a roller-coaster ride of our lives! The ship would actually bend in the middle as we struck the waves, which were about sixty feet high! I mean I really prayed, while expecting at any time for the ship to split in half. The rain came down in sheets, sweeping under the trucks drenching everything in it's path. The deck was awash by both rainwater, and wave splashings. We preferred being on the top deck in case of sinking, so we could get out. Fortunately none of the heavy equipment came loose from their tie-downs, as this would cause a disastrous shifting of weight, and possible capsizing. I sympathized with the Apostle Paul, remembering how he had been shipwrecked, and even floating on the sea after their ship sank. I prayed this wouldn't happen to us, but even if it did in God's providence, we would be kept safe. Bill and I shared much time discussing spiritual things, in light of the situation we were in. I knew where my future would be, even if we lost our lives; at home with my Lord in Heaven!

What a comfort that knowledge is in our times of distress. How I thanked the Lord that I had accepted Christ as my Savior as a young boy, so he could be my indwelling companion in the years to come, and especially on this particular occasion! So many thoughts flood through a person's mind when feeling helpless; when nothing you can do personally will avail.

As a result of the typhoon it took us longer to make our voyage, than the other echelons, and we arrived in sight of Tinian on October 9, 1944. The storm was dying down somewhat, but the waves were still very big. As we approached the coast of our destination island, the waves were striking the 100 ft high cliffs, and splashing twice as high as the cliffs. The side of the island we saw first was nothing but sheer cliffs. But as we sailed around the end we saw the harbor, and a beautifully green scenery in the background, on the hills and valleys, freshly watered by the storm. This was so refreshing, as we hadn't seen grass since we left Hawaii.

The smell of fresh greenery was invitingly invigorating! We were told the island was nine miles long by five miles wide. This was like seeing beautiful farmland, compared to a sandbar, where we had just left.

(Our LST Beaches on Tinian, Our New Home)

The atmosphere seemed to give us a new lease on life, and we looked forward with adventurous anticipation, while wondering what was held in store for us, at this new American outpost. Little did we know at the time, we were being watched at this very instant by hundreds of Japanese soldiers, still in hiding in the hills overlooking our landing spot. It's a fact of history, that after we landed on Tinian, three thousand Japs would still be killed or captured, or surrender on this little plot of ground so recently controlled by the Imperial forces. Of course we were unaware of how many enemies were still out there, in hiding, and waiting for a chance to kill us by one means or another. We would encounter some of them in the days, and months ahead. Isn't it good we don't know the future?

We were taken by truck to the area assigned to our battalion, and given a tent to call home temporarily, while the dozers prepared our Company areas, carved out of the mud left from former sugar cane fields. It rained so much the ground was saturated with water, and as we walked we dragged mud like snowshoes everywhere we went. But the hope of having a new place to live was reviving our souls.

The dozers piled the mud and cane into about a 100 foot square mound, which was a little hollow in the middle, with the tents surrounding it. They would haul this mud away later to fill in hollow places, but for now we thought it was an eyesore. We were pioneering again, yet we knew Seabees would make this place livable in short order. Crushed coral would be trucked in to make solid pads for tents, walkways, and roads. A new coral pit was being dug only about a mile down the road from our camp, from which they would blast into the solid coral mountain, using the material for crushed rock construction wherever needed.

Tinian had been invaded about two month's earlier, sweeping through the extent of the island, but not clearing out the highland portion, with all its caves, and tunnels. There were many places throughout the island in which the enemy could hide, including many sugar cane fields, and hedgerows. Therefore it was a necessity wherever we went to carry our rifles or sidearm 45's, except right in our compound. Our battalion was in a secluded area on top of a hill, and would eventually be like a little city by itself, including our housing, headquarters buildings, dining Quonsets, and various shops. The shops included carpentry, electrician, machinist, blacksmith, plumbing, tire, auto mechanics, engineering, laundry, and several specialties. We were even to have our own ships store, sugar mill, post office, and later a hobby shop for making our own souvenirs during off hours. Of course all of this had to be built and staffed by our own Seabees, so except for the necessities it would require months to complete. We were here to complete airstrips for our new B-29 bombers, as our main assignment.

(The Machine Shop with Enlarged Crew)

The day after our arrival, I was again assigned to the Machine Shop, with Chief Ward in charge. This shop was quite a large steel framed building with a corrugated steel roof, open all around, and as a tool room on one end stood our trailer which on Eniwetok had been our entire shop. We also had an enlarged crew, and shared these facilities with the Blacksmith Shop. The original shop crew was still together, and were to train the newcomers, most of whom had years of machinist experience.

Our chain of command now included several echelons; the Chief in charge of operations; a Warrant Officer over him; and a Master Warrant Officer next; then a Lieutenant, answering to the Commander. They were all good men and were each good friends. The organization was very similar to a private company, with each person performing his responsibility much like in civilian life. I could as easily talk to an officer, as to one of my associate machinists, with great rapport between us.

Upon reporting for duty on my first day back at the shop Chief Ward gave me the day off to explore our surroundings. I visited our airstrip site, and then took a look at the coral quarry, before heading down to Tinian-Town, which was built on the harbor. The main Japanese industrial building was a Sugar Mill, which had been completely destroyed by initial invasion shelling. It would be quite a place to explore!

(The Destroyed Japanese Sugar Mill in Tinian-Town)

Little did I know that the Japanese had a huge network of tunnels going out from under this mill, and extending out to the mountains where most of the remaining Japs were holed up. Our intelligence unit would not discover it for some months in the future. Meanwhile these tunnels provided a means for the Japs to spy on our activities without being detected, moving about wherever they pleased and exiting at various outlets along the way.

(The Author at Tinian-Town Ruins)

The buildings of the town itself were demolished, but where each house had stood, they had what must have been such beautiful Japanese gardens. They usually had fish ponds, and little streams meandering around the yard, with tiny bridges offering a pathway over. Some flowers were blooming. And wandering all around town there were cattle, goats, hogs, chickens, and oxen, all on the loose. I imagine some of these provided food for the Jap soldiers still in hiding.

The natives who lived here had all been housed in a large stockade in the center of the island, which was officially named Camp Churo, but usually called by a less respectful name. The men of this camp were allowed to work in the fields and to fish the waters to help provide food for the camp residents. Aside from that we Americans provided staple foodstuffs.

The kids always begged candy from any passerby's. This camp was enclosed by a high wire fence, and was strictly off-limits to any servicemen. We learned a lot about their customs by observing their toiletry habits, and cooking. Our culture was entirely different, and never has been the same as theirs; lacking any modesty with which we had grown up. Captured Japanese soldiers, or those surrendering over the months, were also housed in this camp, with their civilians, but were not allowed the freedom of going outside the stockade.

We were treated about a week later to a special: our battalion orchestra played for us during our chow time. It was such a pleasant treat! Boy, this was surely different from the customs on our former island! We seemed to be basking in the luxury of a new expanded vision, which was about to change very abruptly. However, a treat still lay before us, for Betty Hutton, the actress was to visit our island and would put on a show up on the highlands, at the Marine camp.

(Betty Hutton Sings for Us)

So we were transported up to the area by trucks, and were delighted by that blond bombshell; a vivacious, raucous beauty, as she went through her singing, dancing, comedian numbers breathlessly. However, right in the midst of her acts loud explosions were taking place just over the cliff behind the stage, with accompanying machine-gun fire.

It so happened the marines were on a raiding party, and had flushed out a cave full of Japs in hiding. This location was still a restricted combat area, and we had been warned of that before making this tour to see the show.

Well, we thoroughly enjoyed the entertainment, but the top military brass was "all shook up" about it. They put a three-month ban on any other U.S.O. shows in this combat zone. So the Marianas Islands had a drought of entertainment for awhile, and as events were to prove, this was a wise decision.

The task before us, as a battalion was to build new airstrips on the west side of the island, but right in the center of the location was a hill of solid coral, and beyond that was a valley. The strip, of course was required to be level, so this meant we must carve the hill down to the same base elevation as the two ends, and fill in the valley. Our engineers did the surveying, staking, and then our demolition men went to work to blast the coral. The idea was to cut the hill off and move the material to fill in the valley. Twenty-four hours a day the crews were busy; at night they used bright flood lights. There was constant blasting, with the familiar cries of: "**Fire in the hole**"!

("Fire in the Hole" Eruption)

It was a dangerous place to be around. Yet the trucks were running back and forth, dozers were moving the debris, scrapers were leveling, grease-monkeys were maintaining the equipment, messengers were running by in jeeps, and all this amid the clouds of blasting coral. One night as the crews were getting ready to go to our midnight chow, they set for one last blast before leaving. My friend, Red Mack, was waiting in the jeep for others of the crew to finish with the blast not far away. Then came the familiar cry of warning, and the explosion followed. **Wham**! A large chunk of coral, about ten inches in diameter hurtled toward the jeep. Not seeing it coming in the dust cloud, Red was struck by the coral right in the chest, killing him instantly!

Waiting in the chow hall for our friends to come in from the strip, we noticed the somber looks on the faces of the men as they arrived, and we asked where Red was, as they all gathered. Then we heard the shocking news! We all lost our appetites. This was indeed a sobering event upon our newfound environment. We had lost another buddy on Eniwetok when Jim Stewart, our fire chief had been killed by the ammunitions barge explosion.

Red was the first Seabee I had met when we boarded the train leaving Los Angeles for Camp Perry, and as we entered service we had been sworn in together in the same ceremony. He had been my bunk buddy on the train, and we had gotten well acquainted during that four day trip. And now he would not be returning home to his wife. It was a very sorrowful parting, but the war still was going on, and we had a job to be done. So we sealed up our emotions and moved on.

During this same period of time, one of my friends who had entered service as an Apprentice Seaman, same as myself, was working on the strip one day when a jeep pulled up along side of him. An officer told him to get in, for they were to report to Headquarters, immediately! Not knowing what the matter was about, he obediently went along, and in bewilderment reported to our Commander. They had his files open, and asked about his ability to read and speak four languages fluently. Yes, he could do that, and demonstrated the ability. They notified him he had been selected by the military Intelligence Agency to be shipped back to the States promptly, for training in Naval Intelligence because of his fluency in languages. We never understood what happened to him from that point on, but at a recent Seabee Reunion, I was updated on his whereabouts. At that time he was a Professor, and Chairman of the Linguistics Department at Brown University! Miracles do happen, and it couldn't have happened to a nicer guy!

Dynamite, for the blasting at the strip was kept in an explosives dump near the end of the strip where the work was being done, and about a mile across the valley from our shop. One night we had just returned to the shop after midnight chow, and about five of us were in the Chiefs office shack, and I was standing next to the screen door. Without warning, a terrific concussion blast hit our shop, and the steel roof raised up and dropped with a shattering clang, and the blast wave simultaneously hit us as we stood in the shack.

We all dived to get under the desk, thinking it was a bombing, and we thought the shop had been hit and debris would be falling on us. Soon we realized it was not a bombing, and we all started to feel our hurts. The whole side of my body that faced the door was numb, and my ear hurt something awful. But I couldn't feel my face, because it was completely numb, like it was frozen. It was scary!

We found out that many men farther from the blast were thrown out of their cots onto the floor as they slept. My eardrum was broken, and it took a week before the feeling returned to my face. The explosion was the dynamite dump, with 128 tons of dynamite which blew up in a single blast! It was caused by a Jap who threw a grenade into a box of dynamite, just as the guard unit was being relieved. Twelve security guards were killed, as one unit of six men was relieving the other unit of six. The Jap had evidently been waiting for that strategic moment to blow the grenade. Many of our Seabees were working near the blast area, but none were seriously injured. The Marine camp that was located about a half mile from the blast was leveled, but I'm unaware of the casualties. What next?

Just three weeks after arriving on Tinian, the Japs had us in mind when the moon was getting toward full. About one o'clock in the morning the air-raid sirens went off, and we were in the midst of a Jap bombing and strafing attack. We still didn't have any foxholes dug because we were living in temporary quarters.

Where could we hide? Some of us sought the refuge of the pile of mud in front of our tent, because it had a hollow area in the center, and unless we got a direct hit the outside berm would shelter us from shrapnel. My heart was in my throat. We essentially were right out in the open with only sky above us. I prayed for safety! The planes were strafing the North field which was five miles away, but other planes were coming at low level across the island all around, and we couldn't tell what they were doing. At least no bombs dropped on us.

Then as I prayed, the Lord seemed to assure me that this was a much bigger island than Eniwetok, and He would keep the planes away from us. My fear settled, and never again, through many air-raids did I have that kind of fear again as I had the faith to believe God would protect me through any danger. During the raid our Marine gunners shot down three of the attackers.

Two nights later, at full moon, the enemy bombers came in again but this time it was at high level, and our anti-aircraft fire drove them off, with no damage done to any of our facilities. The orders to fire against any aircraft had to come from Saipan, which was about ten miles from our position, so they couldn't be sure of what was happening to us locally. Our Marine gunners were supposed to hold their fire until notified by the Saipan lookout to be sure we weren't shooting down one of our own planes. This warning system didn't always work, especially when the enemy attackers came in at low level from the southern direction. And just four miles off our harbor to the South was a Jap held island which acted as a shield from seeing any incoming aircraft if they were low.

Two nights after the full moon, the Japs attacked again, at about 1 o'clock. We were getting jumpy, as the sirens wouldn't go off until after Saipan notified us of an attack, and we didn't trust their early warning. But the planes stayed over Saipan and our North field, bombing and strafing most of the time, but then circling around to our harbor area to our South, they bombed those installations. They had left our area unscathed, so we were relieved. All-clear sirens sounded and we settled back to our work. During this period I was working all night long in the shop, so was awake for these attacks. I felt sorry for the guys sleeping in their tents, especially when we were being attacked and no warning sirens had sounded to get them out of their sacks. At least while working, we could be aware of what was going on around us.

At 3 A.M. a sudden rain storm hit us, accompanied by lightning and thunder. This was a relief, because the enemy couldn't fly in this kind of adverse weather. But lightning storms disrupt radar, so there could not be an effective watch over the skies. In our shop area the lights blazed brightly, lighting up our open buildings, and surrounding landscape. Construction work continued out on the airstrip, with their banks of floodlights. I was working at a lathe near one end of our shop, and facing the road leading past our building, toward the coral pits. I heard a loud shout, and looking up saw one of our men standing in the middle of the road waving his arms wildly, and pointing to the West.

I looked in the direction he was pointing, and found myself looking **face to face with a diving, strafing Zero**, at treetop level, coming straight for us. His guns were blazing, and tracers were marking a path heading right for our brightly lighted shop. We made a beautiful target for him! Someone shouted: "**Douse the lights!**"

I was the farthest one from the main light switch panel, but everyone else was running out of the shop. It had to be me to get the lights off! Many fleeting thoughts fly through a person's mind in a split second, and these thoughts prompted me to act, without any fore-thought. I dashed right toward the Zero, in the direction of the light switch, watching the tracer bullets kicking up the dust in front of me! No time to be frightened, except as I neared the solid 8 inch thick steel surface-plate table, which had about ten inches of clearance to the floor, I thought of trying to dive and squeeze under it. But that wouldn't get the lights off and I knew we were sitting ducks with all those lights on, and who knows how many more planes would be coming. Bullets or no bullets, I did what I knew had to be done! I reached the light panel and pulled the master switch, and we were suddenly plunged into pitch blackness.

Running for a foxhole I saw the Zero zoom past and down the road to the coral pits. He caught them with their lights all on, strafing there, then circling and coming back in our direction, with a hail of bullets bouncing off the coral of the road as he passed us, heading for the airstrip. The Marine gunners had not yet gotten the word from Saipan that it was O.K. to shoot. Assuming one of our own planes wouldn't be strafing us, they tried to get their guns lowered enough to hit the plane. Yet only as he pulled up at the end of the strip did they get him in their sights and they blew him to smithereens. Let me tell you it took a while for us to get our composure, as we realized how close we had come to being shot up. It's good the plane didn't have any bombs.

Imagine; all this had happened while our sleeping tent mates were unaware, because no warning air-raid sirens had gone off. The fact that we could be under attack, without warning, put a scare into us, for the future. Much more was to come, to make us realize we had every right to be frightened. Beware of full moons!

> The Psalmist again comes to mind, as he comforts the believer: "I will say of the Lord; He will save you from the fowlers snare. You will not fear the terror of night, nor the arrow that flies by day,—For He will command His angels concerning you, to guard you in all your ways."
>
> —A VET

(Clay at Entrance to our Battalion Area)

B-24
Forward
Gunner
Volunteer
or
AERIAL
SHENANIGANS

Chapter 10

Forward Gunner Volunteer on a B-24 Mission or (Aerial Shenanigans)

Performing every task to the very best of your ability brings not only self-satisfaction but warrants the thanks of others for whom the service is rendered. And God who sees in secret rewards you openly. We are to render the service as unto the Lord and not unto men.

—A VET

The air raids that had begun at our first full moon became a predictable pattern. However now here on Tinian, where we were much closer to many Japanese air bases, the onslaught became relentless. They even added daylight raids. It was getting to be quite bothersome! Especially at night, the sudden wail of the raid sirens was scary, and since we didn't know how close the enemy planes were we became very jumpy. The timing of their raids was not on any schedule.

So it happened one night at 8 P.M. we were startled from our jobs by the siren that was located just outside our shop. We doused the lights and listened for any telltale sounds. Then we heard the Jap multi-engine planes coming. Their engines were unsynchronized, so they would set up a beat that was very identifiable as Japanese. It sounded like a flight of several planes, and then we heard the high revved engines of fighter planes escorting them.

Surrounding our Seabee camp was a battery of five positions of 105 mm artillery. They opened up with their anti-aircraft fire, and the sound was something to chill the blood. They had the bombers in their search-lights, and the shells were bursting all around the planes. Then we heard the whine of a Jap Zero coming down the airstrip toward our area. We ran into our trailer at the end of the shop, figuring the extra shelter of roof and cupboards would shield us from the flying shrapnel and bullets.

The searchlights had one of the overhead bombers focused, as he started his dive bombing directly toward our shop area. The ack-ack guns were rapidly firing at him as he dove. We braced for the bombs to hit somewhere around us, but just then a blinding flash of light, and explosion as one of the shells hit the diving plane directly, and blew it to pieces. We knew what goes up must come down and sure enough, pieces of the plane and shrapnel came raining down on our steel roof and landing all around the shop. It was a frightening moment of terror, as we expected even larger chunks would penetrate the roof of the shop. But the rain of metal stopped, and the remaining bombers flew away. That night our night fighters downed several of the attackers. Then it was back to shop work again as though nothing had happened. When was the next attack to begin? In a few minutes? Hours? Days? The suspense element kept us jittery!

But I was enjoying the work being assigned me as the shop's precision machinist. Our Chief knew the special jobs that none of the other machinists cared to take on, so these were what I especially liked as a challenge. It was totally unpredictable what sort of jobs would be requested. Our customers came from the Marines, Navy, Air Force, or Seabees. The men from the "ice-cream" barge tied to the pier in our harbor, had brought us a crankshaft, from their ice-cream freezer compressor, that was very badly damaged and needed a miracle to work again. No such parts were available from any source so they begged us to try our best to repair it. It looked like such an impossible job that none of us was about to volunteer to machine it.

Every cam was so gouged with metal that we were sure it would never be useable again. Setting up a lathe to rotate six different cam centers was more of a task than anyone should be asked to do. Even the Chief thought so, and consequently didn't push.

(Our Captured Japanese Lathe)

Well, one night, I was working on the restoration of a captured Japanese lathe which we were trying to put into service. I was experimenting on making standard American threads, that measure in inches, on a lathe that had only metric gears. This lathe was an extra, but was of excellent quality. It was not yet expected to handle any of our work load. So the thought occurred to me, I could clamp the crankshaft project in this lathe, and even if it took a long time it wouldn't matter if the lathe was unusable by others.

So without saying anything to anyone I set up the job in this Jap lathe, and experimented on one of the cams. To my utter amazement, the terrible looking gouged metal was only soft babbitt metal from the over-heated bearing sleeves. Using crocus cloth I was able to clean the cam like new again. I continued with the other cams and in a very short while had the crankshaft looking and measuring like new. Calling the Chief, I had him inspect it, to his dismay also! We were elated! He gave me the job of delivering the part to the barge at the Harbor, and they could hardly believe it.

They were overjoyed, because now they could get back to making ice-cream again, for the various outfits on the island. Our Seabee outfit never received any ice-cream, so the barge officer promised we would get it on a regular basis. He also gave me a ten-gallon container of ice-cream from their freezer to take back to the shop to treat all the guys! We had a party for all nearby shops when I delivered it.

Our outfit had a photography department for official work, and one day the Battalion Photographer arrived. He requested a large "lens shade" for his aerial camera to be made, as he had been assigned the task of taking photos of the various outfits from the air. Before coming into the service I had bought a small lens shade for my camera, so I knew what they were, and their function. Being the only one knowing what it would look like, I was given the job of machining this big one. After sketching the dimensions and mounting requirements on paper, I set out to find the material.

Such a large chunk of aluminum was not available from our parts bin, so I would have to procure it elsewhere. With hacksaw in hand I went exploring in the canefield for the downed Jap airplane I had spotted earlier. The propeller was intact, so I set about the task of sawing off the large round shank portion.

(Lens Shade Material Taken from Wrecked Plane Prop)

Taking it back to the shop, I machined it into the most beautiful lens shade I have ever seen. The photographer when he saw it, was awed! He asked me to come to the photo lab after work hours.

There, I met the Chief in charge of all the battalion photo work, who was a "Life" photographer before going into the Seabees. Anyone who had seen the Life Magazine would realize the credentials of a photographer such as this. It was very impressive to meet such an accomplished person. As thanks for the lens shade, he proceeded to set me up for a special photo session for the picture of "A Seabee."

("A SEABEE"—by LIFE Photographer)

This portrait was in a newly acquired special color film to send back to my parents. The film was 8x10 in size, and when my parents tried to get it printed, Kodak would not print it because such film and the processing was restricted to military use only. The folks would have to wait until after the war to get it printed. This opportunity opened a relationship with the photo lab that resulted in more shop jobs, and favors for me.

Completion of our construction of West Field was reason for a celebration on April 2, 1945. The task of moving the mountain of solid coral and depositing it in the valley in order to make a level airfield was the largest earth moving task undertaken anywhere west of the Hawaiian Islands. My tentmate Trigg and I went out to the strip to watch the first Brigade B-29's land on their home field for the first time. I took pictures of the two huge bombers flying wing tip to wing tip just over the surface of the strip on a flyby. When they landed, the celebrities spoke to the crowd of servicemen, and the Brigade Commander commended our Seabee outfit for such a splendid job. The pilots said the surface of the strip was smooth as a billiard table!

(Opening Ceremonies—B-29 Fly-By)

Ernie Pyle, the famous war correspondent also spoke, congratulating us on completion of a very important job, which would help bring the enemy to their knees. To say the least, we were extremely proud! We inspected the B-29's, and discovered what a mighty warbird they were. They would rain vast destruction on the Japanese war machine, bringing a just conclusion to this horrible war.

During these early days of 1945, color photography was just in it's infancy, so none was available for our use. As a military assignment, our Life photographer was performing some experimental work with color, and needed a new system for automatically developing and processing the color film. During my visits to the lab, he requested me to design and fabricate a system that could process ten inch wide film and paper. Giving me all the parameters, he commissioned me to build one of these "first of its kind."

To obtain the stainless steel required for the tanks, I obtained permission to scavenger in the aircraft salvage yard for any material I could use. This was quite an adventure, going through the cockpits of these planes that had served us so well in combat. Despite their history they now were retired to the scrap heap. I obtained the stainless sheetmetal from the firewall of a plane, as well as other parts to be used. This job took me quite a few hours to shape the tanks, make the scrolling mechanism, and other components, and to test it out. When I delivered it to the photo lab, the Chief was well pleased.

He invited me into the darkroom to observe his testing of the system, and it worked perfectly. He then invited me back on my off hours for training on the processing of black and white film. He also showed me how to build a printer, and even set me up with processing trays, film developing tank, and some chemicals for use in our own darkroom. I had confided to him how we were hoping to build a darkroom on the side of our tent for privately processing our own pictures.

As word was passed on from our photographer about my abilities, the Army Air Force had me do some work for their photo missions, and the work was completed to their great satisfaction. As a mutual service, their photographer invited me over to their facility, and showed me through their photo supply shack.

He said they had stacks of left-over film from their B-29 raids over Japan, which I could have if I could think of a way to use it. Each plane carried canisters of ultra-high speed film which would reel through their cameras while over the target. After each mission the canisters were dismounted, with still ten feet of film left which was unusable in their cameras. So as a gift I received several of the canisters with film, with the promise of as many more as I could use in the future. The supply officer also gave me continual access to all the chemicals I would need as time went on.

It is note-worthy that film was practically impossible to buy in stores back in the States, and we had none available overseas. This supply of film therefore was something special. However, I would need a darkroom to open these canisters, and the film measured 10 inches by 10 feet. So in "Can Do" fashion I designed a film slicer that would produce film in the common sizes used in our cameras. Using the reels, and backing paper that was left over from developing our own films, I was able to produce an adequate supply of new film for myself and my buddies. It required experimentation by trial and error to determine the proper exposure to use on my camera while using such fast film.

With the materials now available for equipping a darkroom, it was time to get the official permission to add this alcove onto our tent. My good friend Chief Driggers was now the Master-at-Arms (MA), and as security chief would be able to get me the permit. Armed with the reference from our Photographer, I was able to convince Driggers that our carpenter, Trigg, and I could set up this tent addition. It wasn't long before Trigg had built the printer, and we had the dark-anteroom ready for service. This was the only such facility permitted in our Battalion. It was only a short time later, when Chief Driggers was able to help me utilize this facility, by a dramatic event.

One night as I had just gotten off duty, I was approaching my tent, and while passing by the Fire Chief's tent I saw Driggers come bursting out, on his way to a fire. We had seen a huge explosion a couple of miles away, down by Tinian Town, and wondered what the resulting fire was. He said the fire alarm resulted from one of the airplane fuel storage tanks that was blown up, probably by a Jap saboteur. Seeing a photo opportunity I inquired if it were possible to ride with him in his jeep to the scene. Running to our nearby tent I grabbed my camera, and joined him in the jeep. We sped the three miles to the fire, and pulled up as close to the flaming tank as possible. He was the first official to arrive on the scene, and being the Security officer he took charge.

He told me to take my pictures quickly before any other officers arrived, and hide my camera under the jeep seat, as they would be the only photos allowed. Because of this permission, I guess mine are the only existing pictures of that fire, unless an official photographer took them later. The heat was intense, and I was afraid the camera lens or film would be ruined by the heat radiating from the flames.

(Aviation Fuel Storage Sabotage)

I later hitched a ride back to our camp, where I could develop the films in our own darkroom. Again I was thankful for another opportunity opened by developed friendships. By the way, I had first met Chief Driggers as we were sworn into service at the same time in Los Angeles, and we traveled together to Camp Peary.

One night a Marine Surgical Doctor came to our shop with a uniquely special request. He carried a thick medical book put out by the American Medical Association; I guess it was the annual compilation of the A.M. Journal. He said he had a ward full of Marines who needed to be "Circumcised." They were in bad shape with infections, and it was most urgent to have a surgical device made that could do the job. Chief Ward said that if anyone could make the device it would be me. So the doctor showed me a picture in the medical journal of the device he was seeking. It was required to be entirely of stainless steel so it could be sterilized. He explained how the device was to clamp, cut, and suture; and the picture showed the various components, but dimensions and details would have to be designed by me. This was a quite a challenge! I liked it!

I started sketching, and soon had my idea of how each part could be machined. Just how to dimension the intimate parts of this device is left to the imagination. We happened to have all the materials, so I set to work, and machined all the parts. Then came the assembly, and all worked well. However, I was not about to personally test the device! Upon completion the Chief delivered it to the Marine surgeon at his base. A few days later the doctor reappeared to report the device had worked perfectly, and cleanly, with no shedding of blood. He brought along a fifth of whisky which he gave to the Chief for the crew, since he found out I didn't drink.

That was not to be the end of the story. A few days later a Navy doctor showed up at the shop, having heard from the Marine doctor how well the circumcising device had worked, and he asked if I could build him an identical unit for his outfit. So I built a second unit. This was still not the end of the story! Our Seabee doctor heard we were supplying the island with these devices, and requested one for his use, so a third unit was built. I felt like going into the business! But that was the end of the surgical project. Now many service veterans can brag about being circumcised by a lowly Seabee machinist's device. It takes all kinds to win a war!

With completion of West Field, and the new asphalt plant built, our men proceeded to cover the two strips with asphalt paving. A large number of B-29 bombers were flown in and parked on individual pads off the taxiways.

(110th Seabee Crew Asphalting "West Field")

They were getting ready for a huge armada of bombers to fire-bomb Tokyo, taking off from Guam, Saipan and Tinian. Planes had been taking off daily for raids over Japan, and it was a daily practice to sit on our jeep and watch the hundreds of planes in preparation, and take-off. We now had four strips at "North Field" and two strips at "West Field."

There were daily mishaps! Occasionally, a plane couldn't lift off under their heavy over-load, and would run off the end of the runway. About two miles directly off the end of "West Field" was the highest mountain, having a cliff with the North end of the mountain about 200 feet lower than the Southern top. The planes would try to turn slightly to the left to be at the lower land elevation when they crossed the mountain. However when loaded their banking must be very shallow, or they would stall and dive into the ground.

Once as we were watching, the plane taking off was unable to gain altitude to clear the mountain, and too heavy to do any banking, so it looked like he would nose directly into the mountain just below the top. However, as he neared the cliff, at the last second the pilot banked the bomber sharply, lifting his right wing over the higher cliff side, missing it, then immediately straightening the plane level. By lifting the one wing momentarily he managed to keep from crashing. I'll tell you that was a thriller!

The night our bombers were leaving for the firebombing of Tokyo was a night to remember! The taxiways were filled nose to tail, nose to tail with planes, and they were taking off from our two strips at about fifteen second intervals.

All pilots were very heroically cautious, that if in trouble on take-off they would try not to clog the strip with wreckage. They would veer to the side, or off the end, so as not to halt the take-off of remaining aircraft. One bomber could not lift off before it reached the end, so he crashed straight ahead, spewing fire bomb napalm along his crash site. What flames! And those fires cannot be put out, so they must burn themselves out. Of course, no one can survive such a crash, as it erupts in flames. All we could say was: "Those poor guys, never had a chance!"

Awhile later, a plane after takeoff from the other strip could not gain altitude. So he dumped his entire load of firebombs, so as not to crash into the mountain, and the bombs landed at the outskirts of the Marine camp. It burned all night long.

History tells us that in that one night of firebombing of Tokyo, the entire city burned, and one hundred thousand people died. A person would think a nation would surrender rather than see further destruction of their people. But, evidently life did not mean more than their effort of saving face, so the war lords kept the war going.

That same night one of our B-29's had engine trouble and turned back to our island. As he approached for touch down something caused him to veer off to the taxiway where a B-24 bomber was preparing for take-off and he landed directly on top of the B-24! Our faithful hero Hugo happened to be in a jeep near the crash site, and braving the flames plunged into the wreckage. He carried two men over his shoulders and got them to safety, then going back in multiple times managed to drag all the men to safety who could not get out themselves. To see this giant of a man you could picture how he was able to carry such loads, and especially with the adrenaline flowing. But the bravado with which he risked explosions, flames and any instant death to save others makes you admire him greatly. As result of that bravery beyond the call of duty, he received a medal, which is in no way recompense for his act of bravery. You can be sure, those men whose lives were saved were thankful!

A Specially Colorful Character

You would have liked to meet a friend I'm about to introduce. Lt. Scranton was a flyer; a man's man, and a friend who would always stand out in memory! He was one of those swashbuckling devil-may-care young men who probably scared the wits out of his flight instructor. He was the Navy pilot of a B-24 bomber, and was nearing completion of his maximum number of flying missions, before his return to the States. We first met him on a whim of his. He was down at the dock where the Navy was unloading a bunch of jeeps and equipment from the ships. He was only site-seeing; not on business, unless it was his usual monkey business! He spotted a jeep that had just been unloaded, with no one around it, so he climbed in, started it, and drove it off.

Successful thus far and with no one tailing him he headed up the main highway that passed our shop area. He came roaring in, raising a cloud of dust. He jumped from the jeep, and approached our Chief. "I've got a jeep here that you can use, Chief, if you will change the serial number for me!" He told what he had just done, and a couple of our guys agreed they would paint another number on it if they could use it. I refused to be any part of it, believing that it was wrong, and also knowing that if caught, it would mean real trouble. Besides that, his arrogant attitude irritated me! Well, such was our introduction to this 21 year old Flight Officer. This was only the first of many visits to our shop. As time went on, I learned to like the guy, and I found out his co-pilot was Lt. Minton who had grown up in school with my brother-in-law, Elton Whildin.

Scranton was an old hot-rodder, knowing all the techniques for modifying a car to go fast. So he oversaw the rebuilding of the jeep engine in the auto shop, with no speed governor on it. It turned out to be the hottest jeep on the island, as no one else dared even remove the governors from their jeeps.

In the process of time we made parts for maintenance of his plane, and other useful things, and he was anxious to do something as a favor for us. We suggested a flight in his B-24. However, he replied that was against regulations, (as if he cared) unless he signed us onto his crew as official gunners on a mission! Well, so be it, we were willing!

It was at this time our forces were in the process of fighting for Iwo Jima. Scranton had been running some bombing missions to help out in this battle, and they were almost to the point where they could land on the captured airstrip and carry out their missions from there. He said the landings would require coming in low over the water, swooping up the cliff and settling onto the strip under the fog. It required some very precise flying! What he did not tell us was that he needed some practice at this kind of gutsy flying.

I hesitate to insert this parenthetical note for fear you will think me one of those UFO nuts, but I swear this is the truth! While the battle for Iwo Jima was underway, one night I was leaving the shop to go to another building. I was suddenly startled to see in the jet black sky a flying object way off in the distance that appeared about the size of a basketball. It was bright red, perfectly round, without any tail trailing it, and it was moving slowly across the sky. I shouted to the other guys to come out to see it. We all watched it progress steadily across the black sky and expected it to crash on the island, but it just continued, evidently way over the horizon, and disappeared from sight.

It must have been a large object at a very great distance away. We were all mystified, and speculated it must be some unknown type of flying weapon. The length of our sighting must have been at least a minute. However, we never heard anything about it from any source, and other people would not believe our story. So I still have no idea what it could have been, although I've often wondered if it could have been one of those UFO's.

(Best Gunnery Crew in the War)

(Seabee Crew—Ranke, Derflinger, Butler, and Rasmussen)

The time came when we were to sign-on to Scranton's flight crew. There were four of us to go from our shop, the remainder of his crew being their regular co-pilot, navigator, radioman, and gunners. I was assigned to take the bombardier's place, manning his gun in the nose of the plane. This was thrilling! The anticipation really had the adrenaline flowing.

Just like the regular missions we lined up on the strip for take-off, and I viewed the runway from way out front, with the engines, wings and everything behind me. I felt alone! Not sure I had made the right decision to take this flight. What would I do if we encountered the enemy? Then came the acceleration, and seeing the blacktop whizzing by just below me. Stuck out in front, being the first one to get there, so to speak. The nose lifted and we were airborne. I had dreamed of just such a feeling. The vibration was something I hadn't anticipated, but the engines were lugging, to lift these tons of plane from the earth. Out over the valley, with the cliff-side mountain coming up. Then he surprised me by banking sharply to the left, a maneuver the other bombers cannot do when loaded!

We climbed rapidly, as I now had Saipan directly in front of me. Sitting in the bombardier's compartment is like sitting in a glass-house. You can see out in every direction except directly behind, and you can look directly down or up, with no obstructions. What a view!

Then he circled around heading for Aguijan, the Jap-held island off the southern tip of Tinian. We could see Japs working in their fields, unbothered by their proximity to our bustling war machine. I wondered if our pilot intended to drop some bombs on these peaceful looking farmers, but I wouldn't know how to do the bombing. And even if he intended for us to use the guns on them, I wouldn't know how, for we had not been taught. But no, Scranton intended to leave these Japs alone. They weren't causing trouble for our troops like the ones on Iwo Jima, where he was intending to go the next day. We cut back across Tinian, and then headed down the straight between Tinian and Saipan, heading East.

By this time I was exploring the compartment to see what I could learn. Ah ha! There's the "Norden Bombsight," the top-secret device that provided accuracy of bombing like no other nation had. If only I could experience sighting through it! I crouched down, and glued my eye to the sight eyepiece. What a sight! Crosshairs and all, with the target right in the center. Never could I dream of such an opportunity, to see through the bomb-sight a target just like our flying men experience on their bombing missions! Out of this world. But now we were coming up on Saipan. I could see the cliffs from which so many Japanese civilians had plunged to their suicide death rather than be taken prisoner by our forces. What useless dying.

As we approached the northern plateau area of the island I saw a bunch of Japanese farmers who were being allowed to harvest their crops in the field, to provide for the prison camp. Some were on a flat-bed truck while others of their crew were loading it from the ground. We were at about 3000 feet when suddenly the pilot put the bomber into a dive. The sound of the engines faded behind me, and I heard the swishing of wind around the canopy. I was alone out in front with the ground rapidly approaching. My heart was in my throat, as we gained speed, heading straight for the truck with workers on the flat-bed. They were watching this plane diving on them, and I imagine it was a reminder of days past. I began to wonder if he could pull the plane up in time. It was getting awfully low for a pull-out! Well if we go in, it will be a fast end for me being right in front. He began to pull out of the dive, and I could picture him straining on the yoke, pulling it back, hoping he had enough altitude to keep from scraping the bottom of the plane on the truck.

What a sight to see the surprised, open-mouthed Japs dive off the truck as I swooped over them just feet above their heads. Wow! What a thrill! I could now laugh to see the startled look on the faces of those men as they dove headlong off the flat-bed. That was quite a sneaky trick he pulled, but it was typical of our friend Scranton! In a little while we were back on the ground, shaking his hand in thanks for the thrill of a lifetime. This was just an ordinary outing for him.

He later came to the shop to tell us he was to be completing his required missions, and heading home to the States, and that we could have "his" jeep. This jeep was actually his officially, and was the second one he had hopped-up, as he had to ditch the first one in the canefield when they had warned of an audit, looking for stolen jeeps. At that time they found 15 abandoned jeeps in the canefields that had been "stolen." They spotted them by use of a helicopter.

Scranton told us that when he returned from his last mission over Iwo Jima he would say goodby by making a pass over our shop like we would never forget.

The next day we were on the alert, and we heard his plane coming. We watched that four-engine bomber skimming over the tree tops at top speed, roaring down the road past our shop. As he neared our main camp entrance he banked that huge bomber into a 60 degree climbing bank and went roaring off, literally into the sunset, never to be seen by us again. What a guy!

> Many are the friends we gain over a lifetime, and each one differs from the other, but they each leave upon our slate a part of themselves; in effect, or in memory. May we each be proud of the tracks we leave in the sand beside those with whom we walk.
>
> —A VET

(Our Seabee Battalion Commemorative Nose Art!)
(The Famous 110th NCB)

(110th Seabee Area Entrance, off Broadway)

(Miss Cora Jacobson, wearing a handmade shell necklace and bracelet from her Seabee)

HIROSHIMA
"LITTLE BOY"
HOW DO WE LOAD
"LITTLE BOY?"

Chapter 11

How Do We Load "Little Boy?"

(Build a Hook to Lift 5-tons) "A Seabee can-do it" ——NO MORE DELAY——

Rumors had been flying for some time now, as we Seabees were seeing so many "civilians" busily at work around the assembly plant we were building for assembling some "mystery object," yet unseen. These civvies were scientists, assigned to watch over the "Project" whose mystery had us all guessing.

Smitty came back to our tent one night, with a fantastic story he had overheard one of the civvies telling. He had said that "if one of our B-29's was flying over the ocean at 20,000 feet, and dropped a device like this 'Mystery Thing,' the plane would be blown out of the sky—even at that altitude." (We immediately "corrected" Smitty that he must have said two thousand feet instead of twenty.) For we all knew that a column of water couldn't reach 20,000 feet!

A verbal order was given us by each of our officers that we were to be aware of a training exercise to be conducted along our main street, "Broadway." This road ran full length through the center of the island; nine miles from Tinian Harbor to North Field. During this exercise all traffic would be cleared along Broadway, and with no cross-traffic allowed. Armed guards stood at all cross streets. The first time this exercise was implemented, we stood outside the shop watching the proceedings. Suddenly we heard a roar approaching on Broadway, and along came a short convoy of covered Army trucks at top speed (governors disconnected). They tore past us heading for North Field. After they were gone, traffic was again allowed to resume. This exercise was repeated several times in the next couple of days. More mysterious activity!

Then, one day my tent mate Ganske was greasing some equipment down at the docks at Tinian Harbor when he heard the dockhands say that the cruiser sitting out in the harbor was the "Indianapolis." It unloaded some cargo, delivering it to the dock under heavily armed guard.

They immediately loaded it into the covered army trucks we had seen in the previous exercises, and the *final* convoy exercise was then activated. Everything was done at amazing speed, and the trucks took off on their 9-mile dash for North Field. Remarks were made that all the armed guards accompanying the cargo were *volunteers* on this mysteriously dangerous mission.

Another tent mate, Kennedy, doing his work at North Field saw the convoy arrive, and deliver their cargo into the new assembly plant. The final preparations at the plant had been the building of a berm all around the plant, evidently to deflect the blast of any accidental explosion.

Another crew had dug a "pit" in the coral bed of one of the bomber parking areas. In curiosity they asked what it was for, and were told it was to place the "Mystery Bomb" in, to prepare it for loading it into the plane's bomb-bay. The plane had been equipped with reversible propellers (first of their kind) to enable it to back over the pit without the nose-wheel interference.

Our tent-mate, Walters, made a habit of listening to his radio while repairing typewriters in his shop at night. He heard "Tokyo

Rose," the notorious Japanese Disk-Jockey/Commentator announcing that the 509th Composite Group had arrived on the island. At the time *we* didn't even know the importance of this Group.

Other preparations were also taking place on the island. Our carpenters were involved in construction of a huge Tent City which was to be a 400-nurse hospital. My friend, Trigg, who was a carpenter, had seen this project and told us it was to house wounded evacuees from our invasion of Kyushu, a southern main island of Japan. More disturbing news was that our battalion was to take part in that invasion, just a couple of month's in the future, and we had hoped to be home by Christmas!

I pondered: "Was there only some way I, personally could be involved in helping end the war before that invasion was necessary?" We all knew the Japanese in their homeland would fight to the last man and woman, and we knew intuitively the casualties would be countless, on both sides. If only God would intervene!

(These thoughts came to me as I pondered the suicide of one of our workers, that had just happened while working on the airstrip. He had purposely laid down in front of the steam-roller he was driving, and let it crush him to death! I had thought that if he wanted to commit suicide, he could have swam over to the adjacent Japanese island of Aguijan, and taken some of those Japs along with him.)

Well, there are many things in life that are beyond our anticipation, or personal control. God works in "mysterious" ways. That reminded me that we had our "Mystery Object" just five miles down the road from us. The time was July 28, 1945.

Upon reporting for work on the night shift, in the Machine Shop, Chief Ward called me into the office shack, and told me he had an important *RUSH* job for me to do. Giving me a sketch, he explained that this "Hook Assembly" had to be strong enough to lift a 5-ton object. It was to be used to convey the "Mystery Object" in our new assembly plant, and in its later deployment. As soon as I was finished fabricating it and he did some welding on it, the two of us would deliver it by Jeep to the 509th Composite Group area.

The job itself didn't scare me, as it was simple enough. Yet shivers ran up and down my spine thinking of delivering it by Jeep, at 2 o'clock in the morning, down the very cliff-side jungle road where four Marines had been ambushed and killed by Japs, just a week before. The only consoling thought was that we had the fastest Jeep on the island, that our expert mechanics had made an actual fireball performer, by hopping it up, with no governor. And we could use it for the trip to North Field.

The job of machining the "Hook" went smoothly, but the tough steel I built the parts out of was extremely hard and slow to machine, so that it took longer to fabricate than what we had estimated. Chief Ward was getting impatient with me about this rush job. Finally, when I finished the machining and assembly, and he finished the welding, the job was completed to our satisfaction.

"Hop in the Jeep, Ras, you're driving," Ward said, as he followed with the "Hook." We pulled out in a cloud of coral dust, and onto Broadway, the headlights gleaming in the crystal clear night air. The convoy we had seen speeding down this road had nothing on us! The five miles straight down Broadway swept by without a word between us. But as we approached the jungle-covered cliff-side part of the drive, we each voiced our concern. Ward said: "You just do as I tell you, and we'll make it O.K."

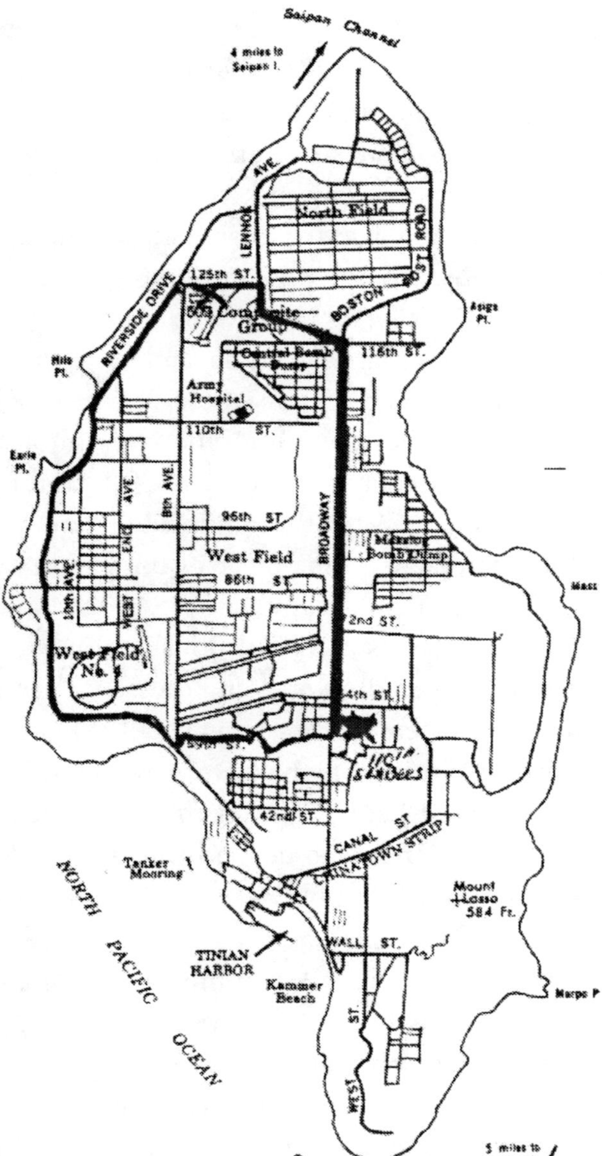

Tinian Island, 1945

(Tinian Map Shows our Route to the 509th)

As we made the sharp left turn heading into the steep downhill jungle ride, the Chief shouted out: “Put it to the floor, and duck down!” What a breathtaking ride that was! The speedometer pegged at the top end while we were still gaining speed. Never, in all my life was I so scared! Not only about the possible gun shots we might hear alongside, but also for the breakneck, crazy speed we were traveling, in a Jeep that was just not designed to go over 60 miles per hour!

As we exited the jungle area at the bottom of the grade we both voiced a sigh of relief. “I told you we would make it O.K., Ras; but you sure gave me the ride of my life! I’ll drive going back!” We approached the tightly secured area of the 509th, and I pulled up in front of the guard tent. We both went inside, Ward carrying the “Hook,” and he approached the officer of the guard. “Password, Sir?” Ward spoke out: “Blank-blank.”

“Sorry, Sir, but that password ended a short time ago.” Chief turned to me and whispered: “Now you know why I was so impatient to get here.” We were really taken aback! What now? Ward insisted: “But we have this part which we were told to deliver personally to the Officer in Charge.” The guard also insisted that without the password we couldn’t go beyond this tent. “Well, get the Officer on the phone and tell him Chief Ward of the Navy Seabees is here with the part he ordered made.” “O.K., Sir.”

We took seats and waited a few minutes until the officer arrived and identified himself. He was sorry for the mix-up on passwords, and thanked us. We watched as he disappeared into the night with the “Hook.”—All this rush; with a job well done; a wild night dash; and “Poof,” it was all over just like that.—Little did we know the giant “Poof” that would atomize that “Hook” just one week after we had our hands on it!

We were quite disappointed not being allowed into the secured area so we could see the object of all the mystery. Now with completion of our job, we decided to take the “High Road,” along the cliffs on the West side of the island for our return to the 110th Seabee area. It was a lot longer route, but we weren't in any hurry now. The night was clear, the moon was shining big and bright, and as we drove along the cliff, I remarked what a romantic sight it was to see the moonbeam glistening off the waters of the Pacific Ocean.

I was thinking romantically because we thought the war was soon to be over, and in my most recent letter to my corresponding sweetheart, Cora, I had told her for the first time that I loved her! What a difference would be our perspective of this scene, if we were again at peace. We had just completed our part of ushering in that peace, and yet we didn't even know it!

Disclosures over the coming years after the war would help us realize the part we had played in bringing about peace again.

The reality of war came upon us abruptly a few nights later, when suddenly at about 1:45 A.M. on the morning of August 6, 1945, all the air raid sirens on the island started to wail. They are really a spine-tingling sound, which we had heard all too often in months gone by. But this was it!

We had been told that even if we had never used a fox hole before, that this time it could mean our life. This was the siren warning we had been told to listen for. We were to get into our fox holes immediately, and not get out until the all-clear was sounded. We did not know,—nor had we been made aware,—that in the event the bomber crashed on takeoff, according to the scientists, even our distance of 5 miles from a possible blast would not permit our survival, unless we were in a fox hole.

The drone of B-29's engines could be heard in the distance, and then silence before the "all-clear" sounded. What it was all about, we did not know! However a few words written to my parents in California just five days before the Hiroshima bombing, did say something of what we knew.

Quote: "Let me tell you, the Army is not disclosing anything that they are about to throw at the Japs, but I know what it is; we helped work on it, and it is terrible. The Japs should quit soon, or they are crazy."

The news rapidly spread around the island on that day of August 6th. An Atomic Bomb had been dropped on Japan, and it had devastated the entire city of Hiroshima! Surely Japan will surrender, we thought! We now knew what the "Mystery Object" was that had been assembled there on Tinian.

Planes again took off to drop leaflets over Japan, urging them to surrender before they would suffer any more such bombings. However, with no response from the Japanese Government, on August 9th another bomb was dropped over Nagasaki.

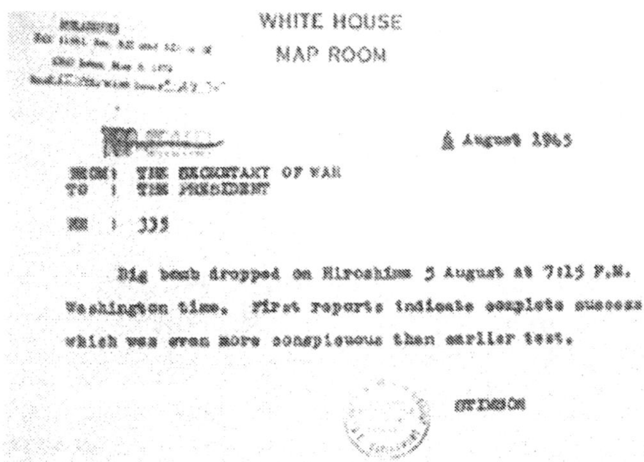

WHITE HOUSE
MAP ROOM

6 August 1945

FROM : THE SECRETARY OF WAR
TO : THE PRESIDENT

NR : 335

Big bomb dropped on Hiroshima 5 August at 7:15 P.M. Washington time. First reports indicate complete success which was even more conspicuous than earlier test.

STIMSON

(Aug. 6th War Memo from Secretary of War, Henry L. Stimson to President Roosevelt)

At dawn, as I was returning from work in the shop, on the morning of the announced, unconditional surrender of Japan, the sky was just then getting bright before sunrise. A row of our tents was silhouetted picturesquely against a lightly clouded sky, so I decided to take a picture of it for posterity. I remarked to a friend: "I will name this: 'The Dawn of Peace.'" And so it was!

("The Dawn of Peace"—Aug. 13, 1945)

Before returning to the States we were forbidden, with penalty of Federal prosecution, from disclosing any information about the atomic bombs. Such as: where they were assembled, point of take-off, or anything technical about the bombs. The entire project was still TOP SECRET, and was to remain that way until the Military chose to disclose it. At discharge I had to sign documents that bound me to that agreement. It wasn't until one-year later, when the newspapers printed disclosures, that I was able even to tell my parents of our activity related to the Atomic Bombs.

Triggering Flashbacks

As the shroud of secrecy, hovering over the WWll Atomic Bombings, began to be declassified years later, I became more aware of my part in the entire scenario,—things unknown to me at the time.

Prior to my entrance into Military service I had worked for Dr. A. O. Beckman at National Technical Laboratories in South Pasadena, CA. During the last part of 1942, and into 1943, I had been asked to make special parts for a secret project that Beckman was personally working on. He would bring me a sketch, and special materials, (one time the material was rods of solid Sterling Silver) and would ask me to make the parts and call him when finished. He would come to the shop, pick up the parts, together with his sketch, thank me, and would then be off on another trip East.

I was made aware that this was part of the "Manhattan Project" only when Dr. Beckman received a Presidential Commendation from President Truman after the war. It was for work he had done on this development of an Atomic Bomb, together with 18 of his employees.

Then again for the last three months before entering service, I was given a three-inch thick book of blueprints for a new instrument Dr. Beckman had been contracted to build for the Government. I was to machine three of every part for three duplicate instruments. This was the very first, commercial "Infra-red Spectrophotometer," later designated the Beckman IR-1. When all the parts were completed, the General Manager, and I assembled the instrument at night and weekends, and he took it to be tested. We packaged it, and a Secret Courier picked it up on a Sunday afternoon to deliver it to its destination; carried with him on the plane. That was the last I heard of it until the mid-1980's. One of the three instruments was used in the engineering to manufacture TNT, and one for making synthetic rubber, since our supply of real rubber had been cut off by the Japanese.

Without disclosing any Classified information, I was led to believe the first "Infra-red Spectrophotometer," as part of the "Alberta Project," was aboard the second B-29 which accompanied the "Enola Gay" on its bombing mission over Hiroshima. It also accompanied the bombing of Nagasaki. They were to use it to analyze aspects of the atomic cloud. The data gathered was used to gain scientific knowledge concerning atomic explosions. I had no knowledge at the time, nor until years later what part I had played in this part of the atomic mission. God doesn't need for us to know what results He is achieving through our service. We must only be faithful, day by day, to do the work He has given us. To Him be the glory and honor.

As a follow-up on this anecdote; after the war I received my education as an Engineering Physicist, and returned to work for Beckman Instruments. I was made their Radiation Specialist on instruments related to measurement of Atomic Radiation, to add to the experience I had already been privileged to partake of in WWll.

It wasn't until I saw a picture of the "Little Boy" Atomic Bomb in a Scientific Journal that I was aware that the "Hook" I had made on Tinian was still attached to the bomb when it was dropped over Hiroshima.

Other information was gathered by research, in a variety of places. It was in this manner that I was given further insight into the project I carried out on Tinian. In the EAA Museum in Oshkosh, WI, they have a full size model of the "Fat Man" atomic Bomb.

Several plaques explain the entire Manhattan Project as it happened during WWll. When the 509th Composite Group was training for the mission in Wendover, Nevada, they practiced with dummy bombs, sized and weighted to simulate the atomic bombs. They had no effective means for lifting it into the B-29 bomb-bay. One time, using a fork lift they dropped it; but it didn't go off. Not wanting such an accident to happen with the real bomb, they tried several different ways to load it, but with no satisfaction. When the Group left for Tinian this problem still had not been resolved.

(Project Silverplate)

Notice what the plaque says about the need for a loading method. "For a time the Manhattan Project was in jeopardy because effective means for loading the weapon into the plane did not exist." **So that was the reason they came to me at such a late date to make the "Hook" which was their solution for loading the bomb!**

(A Truck Lowering the Bomb into the Pit)

A derrick truck would attach to the "Hook" and lower it into the pit. (The bomb was so heavy the front wheels of the truck lifted off the ground, scaring everyone). Then the plane was maneuvered over the pit, and a special winch would be lowered from the bomb bay and attached to the "Hook" on the bomb casing. The bomb would be drawn up into position, not to be unhooked until the bomb was dropped.

I consider it my patriotic privilege having a part to play (small as it was) in helping end the war in this manner. God used my service, even if only an enlisted man, to accomplish His results. Modern day, "politically correct" historians may interpret the "Manhattan Project" in a different light. Those of us who were out there risking our lives against a savage enemy, know that ending the war in this fashion saved millions of lives on each side of the conflict. We were exposed out there to the deadly danger, at that time; and gratefully we returned home to our families and loved ones, and were greeted by a very grateful Nation!

Ephesians 2:10, so simply states what I have found so true in my life, as this account bears witness: "For we are God's workmanship, created in Christ Jesus to do good works, *which God prepared in advance for us to do.*"

—A VET

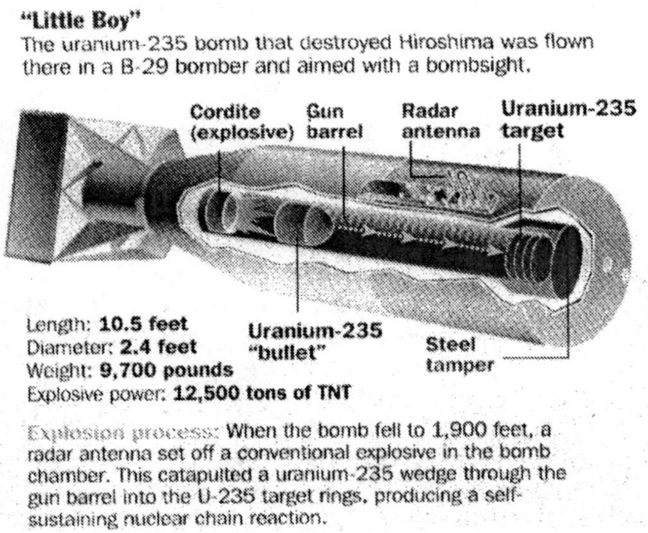

(Anatomy of the Hiroshima Bomb)

Addendum

(B-29's Dropped Leaflets Over Japan Announcing the Apocalypse)

Author's note: "Almost nothing has ever been said to the American public of the warnings given the Japanese people by the dropping of warning leaflets over their cities before the final raids. After the Hiroshima bombing, millions of leaflets were again dropped, warning of more such bombings if the Japanese nation did not surrender."

The above picture and comments below, were copied from "Bombers Over Japan"—by Keith Wheeler, and editors of Time-Life Books.

"On the backside of one of these leaflets it read: 'Attention, Japanese people! Read this carefully as it may save your life or the life of a relative or friend.' It showed the above picture on the frontside of the leaflet, which pictured B-29's dropping incendiary bombs and indicated the cities (circled) scheduled for raids in the next few days. It read: 'We are determined to destroy all of the tools of the military clique, which they are using to prolong this useless war. But unfortunately, bombs have no eyes. So, in accordance with America's well-known humanitarian policies, the American Air Forces, which do not wish to injure innocent people, give you warning to evacuate these cities and save your lives.'

"In the last weeks of the War, the planes dropped blizzards of Japanese-language flyers that, like the Apocalypse, warned of devastation to come, and singled out the cities that could expect to be raided. A message that was printed on the flyers urged people to flee the target cities. Americans hoped they would have the double effect of further disrupting the Japanese war effort and reducing civilian casualties in the raids that were about to begin. The campaign was a psychological success, contributing to the civilian population's feeling of impotence and impending doom. "The leaflets had a great effect on the morale of the people," conceded one Japanese factory manager. "They figured that if the enemy could announce a raid before hand, the enemy was superior."

SUWANNE O
KINAWA
MISTAKE!
PEACE
1945
on Earth
CHRISTMAS

Chapter 12

"Mistakes" Have a Purpose! (Even a Navy Cruise to Okinawa.)

"Circumstances that appear to be 'Mistakes' often have a divine purpose on our behalf. Simply follow God's leading and trust Him, who knows the end from the beginning."

—A VET

After spending 23 months in the South Pacific Theater of war it was such a wonderful feeling to be going home again! The time had seemed endless, and we often wondered if the war would ever end. However, with the dropping of the atomic bombs, Japan was forced to make the decision to capitulate, thus ending the terrible carnage they had begun years before. It required two months following the ceasing of hostilities before the Navy could arrange the transportation for our 110th Seabee Battalion to return home.

During those two months I spent many hours in our library trying to determine a plan for my life as a civilian. With much prayer, and seeking of God's guidance, I was looking at various alternatives. I very definitely was being called of God into some technical field, and one possibility was that I could attend college and study for a profession.

The engineers and scientists that I previously worked with had given me an insight into a very interesting and satisfying professional career. I read books about famous men of science, and inventors. My thoughts always marveled at the feeling of accomplishment these men enjoyed with their endeavors, and the work itself appeared to be as fulfilling as the final results.

So my mind was made up! I must go to college and get a degree in order to enjoy that same satisfaction. I realized that the college education would give me the tools of the profession so I could perform the work. It wouldn't be easy after being out of school almost seven years. But the Seabees credo was "CAN DO!" The field of "Applied Physics," seemed to fit my interest. And maybe eventually, I could even work in the field with Dr. Beckman, my professional hero. That decision became firmly established in my mind. I would pursue it with every ounce of energy God provided.

Communications with my "correspondence sweetheart," Cora, continued with much more focus than in the past. Our letters became more frequent, as I slacked off on writing to other young ladies.

In July, as we began anticipating the end of the war, I had even told Cora: "I love you!" Her response was correspondingly heart warming! She was thrilled when I told her of my plans to attend college, and to pursue a scientific career.

However, in another area of my planning I had a nagging, guilty conscience about an accusation I had made to another high school girlfriend. Even though I hadn't seen her for five years, I had been faithfully corresponding, even for a period while overseas. Because of some suspicious things she had written, I suspected she was having an affair with someone, resulting in an illegitimate baby. I accused her of this in a final letter in which I closed our correspondence. To this she had written an answering letter in which she claimed I had terribly insulted, and deeply hurt an innocent girl. That was in June of 1944, and we had no further correspondence. My conscience still bothered me greatly, and remained a dampening on my relationship with Cora. But God knew what I needed, and He had a fantastic plan of which I was totally oblivious.

The day finally arrived when we were scheduled to board a ship for our trip back to our beloved America. However, on that day a terrific hurricane, with 150 mph winds arrived, sweeping away our tent cabins, and barely allowing us to keep our feet on the ground. We sought refuge in a large Quonset warehouse where we rode out the storm.

(Our Tent Collapsed Only Minutes After
We Left It)

The devastation was horrible. And of course, there would be no shipping out until the 50 foot waves settled down. The harbor was completely demolished, and sand landlocked any ships in there.

We were quartered in the Officers club Quonset for the next three days until we could leave. Our ship was out in open water, still riding out the waves, but we would be transported out to it, by motor launch, from which we would grab onto a landing net and climb up the side of the ship.

With the launch tossing up and down about 20 feet with each wave, it was down-right dangerous grabbing the net. At the top of the wave we scrambled up before the next one brought the launch back up to possibly crush us. However, we didn't lose a single man! But oh, I was so seasick; from smelling the exhaust fumes of the launch, and the rolling of the ship. I laid under a lifeboat for hours before my lungs cleared of the carbon monoxide affect, and my stomach cleared of food, by the wrong way out! But Praise the Lord, we were on our way home!

My Christian buddy Trigg, and I played chess most of every waking hour on our trip back. Then finally came the shout: "Golden Gate bridge straight ahead!"

(Returning to the States, Under the Golden Gate Bridge)

As we entered San Francisco Bay, under the bridge, there wasn't a dry eye in the bunch of us hardened veterans! Oh, how I praised the Lord for bringing us back safe, and sound. We had lost two men in the line of duty during our two years overseas, but we were told some Seabee outfits had lost over two hundred men!

Fireboats sprayed water; an all girls orchestra played, and banners flew on welcoming ships, as a blimp flew overhead towing a "Welcome Home" banner. It was quite an overwhelming occasion! "California here we come!" Oh! What a wonderful feeling!

Once docked on Treasure Island, in the bay, we found out there would be a three-day quarantine period before we would be released for our 30-day leave. Having no wearable dress uniforms, we waited in line for what they called "Survivors Issue."

They re-outfitted us with new uniforms, dress shoes, skivvies, caps, towels and all. Marine shore patrol inspected our duffel bags and seabags for contraband, and declared us "Customs Inspected."

At the end of our three days, we dressed in our uniforms for the first time in two years. We learned how to "square our hat," again, and practiced saluting. (This was something we Seabees didn't do in combat areas). We were all afraid we might not pass the inspection at the guard shack, because we were so rusty at this military dress business. But we made it, and rode the launch to the San Francisco Embarkadero. Once ashore we all had to catch our separate transportation to whichever part of the country we lived. While waiting, I saw three cases within twenty minutes where sailors were short-changed by the cashier girls. We had been warned of that practice going on. That seemed so foreign to the type of people I had always known.

I caught the train to Los Angeles, and settled in for the 13-hour ride. Arriving in Los Angeles I phoned home, and what a welcome I got! My folks hadn't heard from me for the last two weeks, so they were unaware of my whereabouts.

The bus took me to San Dimas, and into the waiting arms of my mother. She met me coming down the sidewalk, with one arm in a sling, from a fall she had, and one eye with a black patch from eye lid surgery. I began to wonder just which of us had been at war! My Dad was away on a short business trip and wouldn't be home for a couple of hours. It was so very good to be home again!

Those two years were so filled with experiences, it would take a lifetime to reveal all.

The experience of sitting in an easy chair in a real home seemed so unreal, like a dream. And the variety of clothes of all colors was a special treat to behold.

My 30-day leave went all too quickly. One week of it I spent working on my brother Marius' sawmill with him. I had invested all my savings, from my service days, in his sawmill business, so I felt like a partner. I never regretted it. It returned more than ten times what I had invested. I visited endless hours with my family, and friends, getting re-acquainted. The days drew shorter before I would have to return to active duty. I still was short several points before I could be discharged. I realized it would be Thanksgiving Day when I must report for duty at the Long Beach Navy base for re-assignment, and my family had planned a big feast for the holiday at our parent's home. I didn't want to miss it! I sent a telegram to the commandant of the Navy base asking him for an extra day of leave. Sure enough, he granted me *until Saturday* to celebrate before returning!

Upon arriving back at the base in Long Beach, I checked in at the re-assignment window. Being Saturday, only a skeleton crew was on duty to sign me back in. I handed my papers to a sailor who seemed new at the job. As he filed my papers in one of two adjacent files, he told me I could find a place to bunk, and would be called up for duty later. Meanwhile we would be given 12-hour passes each day until re-assigned. When I found some of my Seabee buddies, who had returned two days earlier, they all thought I had been A.W.O.L. for two days. When I related the fact that the base Commander had given me extra days to be with my family they would hardly believe it! I explained that I believe in going right to the one who can give the authoritative answer himself, and it worked!

Living so close to the base, I returned home almost every day, making it a practice to keep my clothes at home, and changing there as needed. Some of my friends were already being discharged, because as married men they had 10 points added toward discharge eligibility. Other single fellows were going by train to Camp Parks in Northern California.

It was one of those days, when the surprise bombshell hit! I received my re-assignment orders! "Report for duty aboard the Aircraft Carrier, Suwanne, in *two hours!*" What a shock! I was not in the *Regular* Navy; I was a Seabee! I was trained for land-based duty, not shipboard. All my Navy blue uniforms were at home, with no time to retrieve them, and all I had with me was what I was wearing; my Seabee greens, and clodhopper shoes.

There was not even time to appeal my assignment to get the "mistake" corrected. All I could think of was that dumb sign-in sailor who had put my papers in the wrong file—Regular Navy! I quickly phoned my parents and told them I was boarding a ship, not knowing where it was heading. I had no time to write my sweetheart Cora, to explain why she might not hear from me for awhile. I had suddenly been thrown into a desperate state of confusion!

As ordered, I boarded the Carrier with a few other new assignees, and for the first time found out what it was like to set up my home-base on a ship, surrounded by Regular Navy sailors. Most of these men were veterans, and survivors of the vicious Kamikaze attacks off Okinawa, which had seriously damaged the ship, sending them back stateside for repairs. The war time crew usually consisted of about three times the number of men as they had now. In peace time it was now being operated, short-handed, with each man doing double duty. I urgently tried to convince the Chief Engineering officer to give me a short liberty to get my clothes from home, but he absolutely refused, not believing my tale of woe. (It was most likely God who kept me aboard ship, because I had vowed in my mind not to return to the ship if I once got off.)

I was assigned quarters in an Engineering compartment, adjacent to the Machine Shop. There were 24 of us in the compartment. Our bunk beds folded up in daytime hours and were allowed down only at night. My bunk was the lower of three by the outer ship's bulkhead. I even had a porthole to look out of, and found it was just above water line.

Our first task was to unload missiles and bombs from the ammunition storage. We loaded them onto barges which carried them away. This task took us two exhaustive days. Then we were given our orders. We were embarking for Okinawa, to bring back a load of soldiers, who had fought so valiantly in our last major battle of the war. But at this point I struggled to feel patriotic. I had just returned home myself, and now I must go back out there? I just could not fathom such a drastic "mistake"; and no one would even listen to my complaint about being a Seabee, wrongly assigned to a ship.

The first light of day came when the Engineering officer found out I was an experienced machinist, and he needed a man to take charge of the Machine Shop, with its crew. I got the job! First, I must clean out the "Steering Engine" room of all the diesel fuel sloshing around on the deck since it had spilled during the last storm they rode out. (The one that had delayed our leaving Tinian). We were soon out of sight of land heading West again across the wide Pacific.

Sure enough, we hit another storm that was to last several days. The "Steering Engine" room is in the rear of the ship, just forward of the driving screws, and rudder. In the big waves this part of the ship makes a figure "eight" as it swings and sways up and down. It really feels like an elevator that doesn't know if it is to go up, down, or sideways. That motion, combined with the smell of diesel fuel kept me busy filling my pail with a combination of mopped-up fuel, and what came back up from my stomach. This job lasted three days, and by then I was an old "salt," enjoying the motion, although I'm sure I was pounds lighter.

My first day's activity in the Machine Shop consisted of determining the shop inventory of parts and materials. I found that in the Regular Navy our shop also provided all the parts for the plumbers. I soon learned what all the pipe fittings were for, and labeled the parts bins for my own benefit. Once that was completed I found the Machine Shop to be beautifully equipped with better machines than I had seen since leaving my civilian job. I started to feel at home then in the shop.

But one day, after working a busy eight hours in the shop, I was enlightened to the fact I also had to stand 4-hour watches each day in addition. "Lights out" were at 10 P.M. and my first watch was from midnight to 4 A.M. No one was allowed in their bunks after 5:30 A.M., so I didn't get much sleep. The watches rotated daily.

My watch post was on the hanger deck, where every half-hour I must climb outside the ships hull, and up the steel rungs on the side of the ship to the Whale-boat. There I was to start the engine, and warm it up. The purpose was to have the boat ready in case of a "man overboard" emergency. I hoped I wouldn't be the first casualty! As time went on the stormy waters got worse. Then came the night when the waves were crashing against the side of the ship, reaching far above the hanger deck, even up to the Whale-boat. It was progressively getting worse as I came on watch that night. I climbed up to the boat once, and was almost swept off the rungs by the waves as I climbed. Once back aboard the ship I vowed I would not do it again. The next time it was due I saw the waves were beating against the bottom of the Whale-boat, so I told the Chief I wouldn't go out there again. I felt it wasn't worth the risk to my life to have the rescue-boat ready.

The Chief accused me of "thinking for myself." That is not the Navy way! I absolutely refused to go out, and so was told to call the Master-at-Arms officer, and tell him. No sailor ever does what I was threatening, without being accused of mutiny! The officer responded to my call and came down in a hurry to see what was going on. When he saw the crushing waves he told me I had made a wise decision, and he would back me up. He ordered me not to go outside the ship, but to suspend the operation until the waves died down. My watch ended, and I climbed into my bunk. Next day I found out the waves had beat the Whale-boat to pieces, destroying it. The Lord was watching over me! Even in the "Regular" Navy with its firm regulations.

Now that we were well under way, I still had the problem of not having any Navy blues, dungarees, chambray shirts, and shiny dress shoes. The Captain's inspection was coming up soon, and before I could get new clothes, an inspection was called, to muster on the Flight deck. My officer gave me orders to hide out in the shop's tool crib. He couldn't have one of his men be seen in Seabee greens and suede boots. So I got past that inspection! Before it occurred again I had bought an entire wardrobe of sailor blues. (I felt bad, knowing I had an all-new outfit lying unused at home, and I now had to pay for these.)

One Sunday, while things were slack, I went up on the Flight deck. It was a stormy, cloudy day, and the deck was completely empty. I looked out on the gray sea in all directions, not seeing a sign of life—nothing but white capped waves, clear to the horizon. I have never seen a more bleak scenery; as though I didn't already feel lonely! My thoughts were a few thousand miles away, with family and my Cora, who would be wondering why this sailor boy was not writing, now that he was back stateside. Little did she know I was at that very moment heading away from her, unable to see her as we had hoped. It was good that I was busy most of the time, taking my mind off myself, and my situation. I was in a very strange environment, among men accustomed to strict military protocol, separated from family and friends, among strangers. And I was heading away from my sweetheart who I had expected to see very soon, and she would be expecting my letters, with none arriving. What would she think? I was in a very deep state of depression, especially during my off hours. This was a foreign thought pattern for me, as I was usually very optimistic. I found consolation in reading God's Word, while sitting on the side of my bunk, and praying. I was consoled by the knowledge that the Lord had said He would never leave me or forsake me. This even brought opportunity to share with the other men my faith in Jesus Christ, as they couldn't help but see me, and to answer their questions. I was then directed by them to other Christians with whom I was able to fellowship.

Unexpectedly, one day came an urgent call from the ship Captain's office, for me to report, immediately! Now what? As I entered his office which was in the Operations Tower above the Flight deck, I felt really out of place. Here was an enlisted veteran Seabee, standing at attention before the commanding officer of this victorious Navy fighting ship, on strange territory. But for some reason I had no fear! What more could they do to me? A land-lubber aboard an ocean-going ship; wasn't that enough?

"Seabee!"—He paused as though to let that word sink in. "What, in God's name are you doing aboard *my* ship?" Boy, I was waiting for this! Now's my chance, I thought. "Sir; that's what I've been asking my officers since I came aboard; Sir!" "Well, I have your papers here in my hand, but I have no complement of Seabees allowed on my ship! I don't know how this mistake happened, but I intend to correct it! So I'll give you a choice of two alternatives. (1) I'll give you a raise in rank and swear you into the Regular Navy. Or (2) I'll drop you off at the first arrival port back in the States. Take your pick." Without hesitation I replied: "If you don't mind, Sir, I'll take the second alternative." "So be it!" Then he added: "Even though we would ordinarily have a Chief doing your job in charge of the shop, your officer tells me you're doing fine work for him in the Machine Shop. Keep up the good work as long as you're with us. Dismissed!"

God was expediting my discharge, with a good record. I was collecting double discharge points for every month aboard ship.

The ship sailed on, at a very slow rate, because of the stormy seas. However, now I at least had some hope for the future, to be put ashore again. After almost two weeks, we now were passing the Mariana's Islands where we Seabees had so recently been doing our duty to end the war. Traveling at rates as slow as four knots, was a very boring and monotonous way to cross this vast ocean expanse. But the various machining jobs required in the shop, and my separate duty hours, managed to put the miles to the time. Finally after three whole weeks we steamed into Naha harbor, at Okinawa. Our ship anchored in the harbor a couple of miles off shore, next to our sister ship, carrier. We were to be there until they loaded about a thousand soldiers, to be housed on cots on the hanger deck.

Now, I was to learn what it means to the support functions when the ship is sitting idle in a port. This is the time maintenance can be performed on any machinery, or equipment. The projects started to pour into the Machine Shop almost immediately. Each of we machinists were busy on one machine or another. But off-hour jobs fell to me as machinist in charge. After my first eight-hour duty, I was to be on call. Sure enough, after evening chow, I was hoping to have some time to write letters, but an officer appeared with a machine job. It was needed during the night shift, and so I returned to the shop for another 3-hour stint.

At midnight I returned to the compartment for what I hoped would be a quiet night of sleep. It was one A.M. when I felt someone shaking me to wake up. What did he want? A rush job, needed again for the night maintenance crew. Back to the shop, for another few hours.

(Picture Taken from the Suwanne, Looking Over
Our Sister Carrier Toward Naha, Okinawa)

As soon as the repaired part was delivered to the crew, I hit the sack again. But 5:30 A.M. came in half an hour, and with that the boatswains whistle to hit the deck, for morning wake up. Bunks had to be made up, and slung; meaning no sleeping in them until off-duty hours in the evening, after chow.

After breakfast, I was called to the shop an hour before my shift was to begin. Another rush job needed right now! Then came my regular eight hours on the job, with work arriving continually from all departments. Chow down, and back to the compartment to begin what was to be a continuation of what had occurred the night before. For the three days we were in the harbor I had about 2–3 hours sleep per night. Nearly exhausted, I was living for the hour we would pull out for our voyage back Stateside.

With the last batch of soldiers loaded aboard, we got the word we were to be underway. No sooner had we cleared the harbor than I sneaked off to the parachute loft, which was out of the way of regular traffic. There I found stacks of mattresses in one corner, and I climbed up about twenty feet, finding a spot behind the mattresses where I couldn't be seen, and laid down and immediately fell asleep. It wasn't until twelve hours later that I woke up. But oh, I felt so much better for the sleep! I slid down from my perch, and to the head (restroom) to freshen up. No one was concerned about my being missing, for the guys suspected I had hidden out to get some sleep. Even my officer didn't question me, but looked as though he also knew what I had been up to, for he had personally been responsible for some of my extra hours.

We had arrived in Okinawa on December 20, 1945, and three days later were on our way back to the states. As Christmas neared, the mood on the ship began to take on the flavor of the Holidays. Even the cooks seemed festive, and prepared some special dishes.

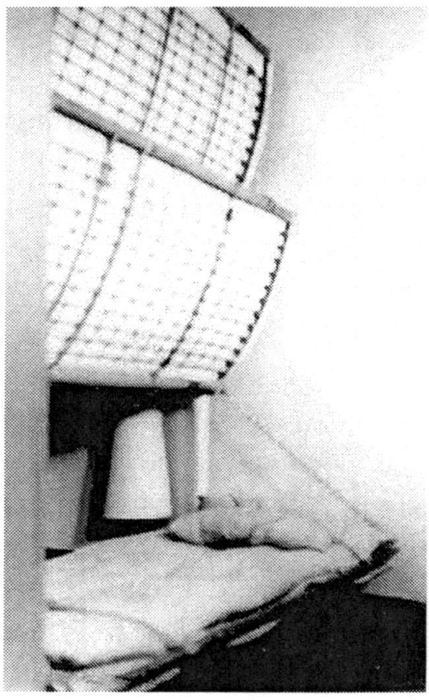

(Our Bunks, Bottom One Mine)

I had more time to explore the ship, from stem to stern, port to starboard. But the hanger deck was getting to be a place to avoid. With a thousand soldiers, not used to stormy seas, the stench was overwhelming. Most of them were seasick! Poor guys; I knew just how they felt, and they looked as bad as they felt! After the battles they had experienced on Okinawa, this was something else! But they too, now had the hope of getting back Stateside.

One Sunday, with time off regular duties, a bunch of the sailors were up on the Flight deck playing football. Now this is a real challenge. There are no fences to keep the ball from going overboard, and with the pitching of the ship, it was somewhat difficult to keep your equilibrium. From time to time you might be running uphill, and the next time going in the same direction you would run downhill. It was such a time as this when the ship changed slope as one of the guys was making an end run. With the energy required to run uphill, he suddenly found himself running downhill, right for the edge of the deck, at top speed. One of the fellows on his team thought in a split second and tackled him just as he pitched over the side. Holding him by the legs, the other fellows managed to grab him and pull him back up to safety! What a narrow escape! But the game went on, even though one of their number had nearly fallen 100-feet to the stormy waters below.

Christmas morning arrived and there was quite a spirit of anticipation among all the crew. We were to have most of the day off, except for shortened hours for the men needed to keep the ship moving, on course. We were even allowed to sleep in until eight. As recorded in a letter I wrote to Cora, at noon we had a Christmas party on the hanger deck in which we sang Christmas carols. We then had several talks by various officers and followed by distribution of packages to all crew members. There were gifts from different parts of the country, and my box was from Westwood, California. They were a very pleasant surprise and everyone was delighted with what they received. In my box was a lot of candy, cookies, a chess and checker game, comb, stationary, a sewing kit, Christmas card, soap, shaving cream, tooth brush, etc. I certainly appreciated it, and the time and effort it took to put it up, by the volunteer. I immediately wrote a letter of thanks to her; it had meant so much to me. Even now as I write this it brings tears to my eyes, to realize how much the people back home appreciated what we servicemen were doing.

At 2 o'clock in the afternoon, we were served dinner. It consisted of chicken-rice soup, turkey, apple salad, cranberry sauce, peas, sweet potatoes and gravy, mince-meat pie, ice-cream, candy, coffee and bread and butter. Boy, it sure tasted good! Christmas day we passed within sight of Haha Jima, in the Bonin Islands. That's one of the islands the Japanese planes were flying from when they were bombing us on Tinian a year before. That night they showed a movie on the hanger deck, called "Winter Time," starring Sonja Henie, ice-skating.

The next day, they called us up onto the Flight deck for a talk and fresh air and sunshine, while they searched the ship for 20 gallons of liquor that was stolen the night before.

God's plan for me on this cruise was now about to unfold. He makes no "mistakes," as I was about to find out by a surprise event! Our Engineering Division had a laundry which washed all our clothes. One day the main washing machine broke down, with no parts to repair it. With Seabee "can-do" spirit I rigged up a make-shift washer, using our Shaper machine for power and the cam action, and a G. I. can, similar to our windmill washers we built on Eniwetok. We were in laundry business.

One of the sailors I had never met before this, came in to the shop to do his washing, because the word was getting around about our washer. In the course of waiting for his washing, he and I started comparing notes, and I found he was from Pennsylvania, near where I had grown up. If you recall I had broken off correspondence with an old girl friend from Pennsylvania back in 1944.

as I accused her of having an affair with someone. I was still haunted by the thought I had maybe accused an innocent girl.

Well, in our discussion we started inquiring about girls we each knew from Pennsylvania. We each pulled out a picture of the same girl; same name and everything about her. Then he said, he had "shacked-up" with her for two weeks, and she had a baby by him, out of wedlock. He said she hadn't married and told me where she was now living. The time of the baby's birth coincided with the time I had accused her, by letter. My emotions were mixed! In the power of the flesh I could have thrown this shameless guy overboard from the top of the flight deck, with no remorse! But God gave me a sudden peace, by this answer to my accusation of a girl who had betrayed my affections in such a shameful way. I lost my guilty conscience! The way was now completely clear for an ongoing relationship with Cora, if God so led. I suddenly realized why God had sent me on this ship, at this time, and His bringing this sailor and me together. He had worked out all the tiniest details, "working in mysterious ways, His wonders to perform." There was no other way that I could have met this man, to obtain the knowledge of his affair, except being on this ship with him.

This memorable voyage ended with a real bang. We struck an extremely severe storm which was lashing the California coast. As a result, San Francisco Port Authority refused to let us come into their harbor, so our Captain changed course for Seattle. As is the custom on a ship, when a course change is about to take place, that could possibly mean a severe listing of the ship, they give a ten-minute warning over the loud speaker.

It so happened that they were serving chow at the time of the course change. I had just gotten my tray full of food and sat down at the long table. The bench you sit on is attached to the table. Unknown to most of us the maintenance crew had unclamped all the tables from the deck, in order to polish the deck in preparation for our docking at the end of our voyage.

As the ship changed course by 90 degrees, to head North, the ship heeled way over. Amidst much shouting, all the tables started sliding on the steel deck. Starting out slowly, they gained speed as though in formation they all headed toward the opposite bulkhead. Of course, we sailors were along for the ride. As the tables slammed against the wall, one by one piling on top of the next, the tables stopped but our trays full of food didn't stop. Everyone on the wrong side were served the food in their laps. But this was not the end. As the ship heeled in the opposite direction, so did the tables. By this time the deck was well greased with food, so the trip went faster to the other wall, we all riding with them. What a mess! Everyone was doused with food and drink. By the time the ship settled into a stable condition, amid much confusion, we began to gather our wits, and to venture up from our benches, walking and slithering on food-goo all over the place. What a way to spend your last night at sea.

But the night had only begun! As the evening got later, so did the storm get stronger. Being a land-lubber by nature, and trade, I began to despair of our reaching port safely. My mind went back to the experiences of Paul, the apostle, who realizing when we are in the care and keeping of almighty God, nothing out of His will can take place. So I entrusted my keeping in the hands of the Lord, and even though I was greatly concerned for safety, I trusted Him. When we climbed into our bunks, we had to put our safety straps on to keep from rolling out onto the deck. I prayed! Continually! And I planned what I would do if the ship capsized. How would I climb up the ladder if it was on its side? All kinds of thoughts! I guess I slept some, but I had to stay awake to pray, I thought. Whatever happens to our faith in duress situations like this? I confessed my weak faith to the Lord, and asked Him to forgive me. "Lord, I believe, Help thou my unbelief!"

By daybreak, the storm had subsided somewhat, and we were all much relieved. I was called to the battle-station where I was assigned for emergencies, and shown the "List recording," which records the amount of list the ship is taking. During the night the ship had taken a list of 29 degrees. It was designed to take a maximum list of 27 degrees without capsizing! As the ship had slid down the huge wave, and listed 29 degrees, it had reached the trough of the wave and the next wave had knocked it upright again! God certainly watches over His own children! Observers said the Flight deck had sliced through the water, as the ship heeled over!

Shortly after dawn, January 10, 1946, our ship turned into the Puget Sound for it's ninety mile trip down the sound to the Seattle harbor. What a beautiful trip that was, in perfectly calm waters. It brought to mind that wonderful promise we have from the Word of God. After the time of testing He brings us into the calm and peace of the lush green valley of life, safe in the care of the Good Shepherd. This was to be the port of my departure, as the Captain had promised.

> "No trial is too difficult, no hill too steep, no valley too deep for the Good Shepherd to bring his sheep safely to the shelter of his fold."
>
> —A VET

(Clayton on Board the Suwanne, in Seattle Harbor)

Peace Dividends

or

New Beginnings

Chapter 13

Peace Dividends!

> Scripture says in Philippians 1:6: "Being confident of this very thing, that He who has begun a good work in you will complete it until the day of Jesus Christ." God had begun a work in this Seabee that he would see to completion, not only in his military experience, but through the rest of his life! Of this I am confident.
>
> —A VET

Arrival in the Seattle harbor was the beginning of the last chapter of my life as a Navy man. The Captain was willing to keep his word and put me ashore, as promised, but he gave an alternative. Since I was so close to having enough points toward discharge, they agreed I could remain aboard ship, while receiving the double number of points each month, as though oversees. The hitch was that soon the Captain was expecting orders to cruise to New York through the Panama Canal, and if I was still aboard I was required to cruise along with the ship. It seemed like a worthwhile gamble, so I agreed to remain shipboard. In expectation of my discharge, I was given the assignment of training a new man to take charge of the Machine Shop.

(Seattle, From the Suwanne Flight Deck)

By this arrangement, the new man and I were able to alternate days on duty in the shop, and have 24-hour liberties every other day. This was really an ideal situation, and we each enjoyed it.

Cora, meanwhile had been watching the newspaper each day and saw our ship, the Suwanne, listed as having arrived in Seattle on January 10, 1946. She was elated! She had earlier gotten a letter from my mother explaining my unexpected cruise on the Carrier. I also had sent her a letter from Okinawa.

On my first day of liberty ashore, I waited in line for a pay-phone to make a call to my sweetheart in Raymond, Wisconsin. I finally got through, and had Cora on the line. It was a fairly good connection at first, but being a "party line," we heard about six clicks indicating that many other inquisitive people were now listening to the conversation. We were somewhat restricted, and feeling awkward therefore in what we could say to each other, except cordial dialog. But it was a wonderful experience to talk to this one with whom I had been corresponding for two long years! We promised to keep the letters coming, and I would let her know the outcome of my quest for discharge into civilian life!

The word from the party line spread rapidly around Wisconsin that Miss Cora Jacobson's fiancee was back from overseas, and could be expected to be making a visit to this farm community. I'm sure the gossip included speculation of marriage, and all that preceeds it!

During that first liberty, it was my first-time opportunity to see Seattle. I fully expected the people to be sick of sailors, since this had been a liberty town for the Navy for so many years. However when I entered a nice restaurant for dinner, I was treated to a very unexpected cordiality. Being alone, I sat at the counter, and was looking at the menu when a well dressed businessman came and took a seat alongside me. He introduced himself, with a friendly bit of conversation, and looked at his menu. Then he asked me, what my favorite specialty was on the menu, since this was civilian fare, and that I must have been dreaming about something really extra special to have, since just arriving Stateside.

Then he surprised me by saying: "Take your choice of the best on the menu, for this meal is on me, in thanks for a job well done!" This was a favor, which was entirely unexpected. In amazement, I responded in thanks, and for the remainder of the meal, we conversed as though we had been friends for life. This episode changed my attitude entirely about this liberty town. Even today I have a warm feeling about that city!

Liberty meant roller skating most of the time at their beautiful roller drome. The young people were so friendly, and the girls wanted to skate with me so much that I began to have guilty feelings, because I knew my own girlfriend would be jealous if she could witness it. Yet I knew that if Cora was there, none of the other girls would have a chance!

One evening, I was whiling away the time, expecting to return to the ship for my night's lodging. While window-shopping, a young man approached me with a handout sheet, and invited me to the Civic Auditorium for the night to hear a very special Christian program. This was just what I needed, so of course I was there when the doors opened. It turned out to be a terrific meeting of song and inspirational message by the world renowned Missionary, Hubert Mitchell, a good friend of my Dad from Los Angeles. After being driven out of the mission in Borneo by the Japanese, he returned to the States and managed the Union Rescue Mission on Los Angeles Skid Row. He had lost his wife on the mission field, and only a few month's previously had remarried, and Mother and Dad had attended their wedding at the Church of the Open Door (my home church). Mother had written me *five pages* of a letter describing the ceremony, and the beautiful bride.

The musicians for this night were the Gospel Quartet from Dr. Charles Fuller's "Old Fashioned Revival Hour." It was led by our friend Rudy Atwood, the fabulous piano player for the broadcast and for the quartet, and who was a long time friend of my brother Alvin. (In fact, Rudy and Alvin had played a piano duet in the Hollywood Bowl at a concert some years before.)

Possibly you can imagine the thrill this meeting was for me after being overseas for over two years, encompassed by ravages of war. I can't describe the depths of spiritual emotions the music and message was to my heart, which was literally starved for spiritual things. Afterwards I went up and greeted Rudy who was really surprised to see me there, and he asked to be brought up on Alvin's activities. Then I went to Hubert Mitchell, and renewed acquaintance with him, and told him about Mother's letter describing their wedding.

He took me by the arm and led me over to his wife, who was surrounded by many women admirers. He interrupted her to say: "Honey, I want you to meet the son of my very best friend, Mr. Rasmussen." He wanted me to describe the letter Mother had written, and they both were thrilled like newlyweds! I was certainly pleased to be treated so royally. It was a night to remember, and it was a rekindling of my spiritual enthusiasm.

On another liberty, I had planned to take a bus up to the famous Snoqualmie Pass Ski Resort, which was in the midst of the deepest snow season. It was only a couple of hours ride East of

Seattle, and we were there. That was my first view of a real Ski haven, and I was impressed, having dreamed while overseas of taking up skiing when home again. After a day of picture taking and watching skiers, it was time to head back to Seattle. The crowd waiting for the one bus due through there, was overwhelming. When the bus arrived it was already stuffed full without even standing room left.

(A Day at the Snoqualmie Pass Ski Resort)

So, some of those stranded sought rooms for the night but I was unable to do that since I was due back at the ship. As you can expect, my heart sank; but remembering I was experienced at hitchhiking, and being a sailor, thought it would be a cinch. After standing along the road, with my thumb out, I was passed by every car, and yet I could see some of the cars would stop and pick up girls along the way. It so happened two girls came along and invited me to hitchhike with them, and as soon as the first car came along we were picked up by a man going all the way to Seattle.

The man was a wild driver, and once on the icy, snowy road at a long straightaway, he decided to pass the long string of cars ahead. Hitting 70 miles per hour we proceeded down the wrong side of the road for a good half mile, and at the end we saw a car coming up the hill headed straight for us. He swung the car in front of the first car in line, and we slithered back and forth on the ice, but he was able to get it under control without going off the embankment! What a scare. It reminded me of our wild ride on Tinian, on the night we delivered the "Hook" for the A-bomb! Well, without any more scares we arrived in the outskirts of the city. They were all hungry, so we stopped at a restaurant, and they insisted on buying my dinner. After being let off at the bus terminal, I caught a ride out to the dock where I had left the Suwanne.

Surprise! There was no ship there! Now what? It was nowhere to be seen, so all I could think was that they had sailed for New York, without me. I found the Port Captain, and asked about the Suwanne and was told they had sailed up to Everett, and would be berthed there.

I then caught a bus that took me the seven miles North, thinking all the time they had probably declared me a deserter. But to my pleasant surprise, they made no fuss over my absence, and I was admitted aboard without any problem. What a relief!

(Last View from the Suwanne)

(Suwanne CVE-27)

The next day, about twenty of us were called to the hanger deck, and told our notification of discharge proceedings had come through. We were to be put ashore to report to the Separation Center, to await transportation to our point of discharge. My discharge would be in Long Beach, California.

What a thrill this news was! However, in the back of each man's mind was the thought of being called back because of some technicality, and being required to serve longer. I knew the Regular Navy Machinists were ineligible for discharge, because of shortage of personnel, and had been "frozen" to their jobs. I belonged to the Reserves, USNR, so that is why I could be discharged. Of course the Navy could change their minds, and that is what I feared. Another of the men had served only nine months total, and we knew a mistake had been made in mustering him out of service, so he had real reason for being fearful of it being discovered.

It was with mixed emotions I bid farewell to my newfound friends, thinking how little time it takes to bind men together in our dangerous walk of life. Yet we were elated to be getting off the ship, and I truly did not want to go back aboard for any reason.

I forgot to take my pillow with me, so during the next few nights I learned to sleep pillowless. Everyone at the Separation

Center was anticipating getting home, by the quickest means possible. They were drawing lots each day as to who would get to fly in a plane to their destination. My number never came up, so after two weeks of waiting anxiously, our train arrived to take some of us to the Los Angeles area.

Several of us from the Suwanne were going to Long Beach, so we rode in the same railroad car. This car was equipped only with upright coach seats, so there was no place to sleep otherwise. It was a long trip South, and it soon became very uncomfortable trying to sleep sitting up, so I climbed under the seat and went to sleep on the floor. During the night we were traveling into the mountains of Oregon, and all at once the train came to a stop. I got up and went to the rear of the train, out onto the observation platform, and the scene was fantastic. It was bright moonlight, and we were surrounded by deep snow, and the fir trees all around us were hanging heavy with fluffy snow. The train was stopped while it waited for a train snow plow to clear the tracks. It was very cold outside but nice and comfortable inside the train, so we went back for some more sleep. One of the men was of communist or socialistic leanings, and we spent many hours arguing his philosophy versus ours. It whiled away the time, but we were unable to budge him away from his socialistic ideas.

We finally arrived at the city of Long Beach, and were bussed to the Separation Center at the Long Beach Naval Base. Then started a battery of testing: complete physical, psychological testing, debriefings, and finally the paper signing. All along the way at each waiting line, men would come out and read off a bunch of names, that indicated questionable situations. We each expected to have our name called saying we couldn't be discharged. Man, what anxiety, and stressful tension! So close to discharge and yet seemingly so far! Finally the roster of dischargees was complete and we were told the date of our graduation into *civilian life.* I called my folks and gave them the time and place of our discharge ceremonies, and they promised to come. At the appointed time, Mother and Dad arrived and then my brother Lester and Jean, with son Dean, and we sat together while several Navy Officers gave brief speeches, and a recruiting officer tried to get us to re-enlist. Yes some of the men did! Then as our names were called we marched up to the platform to receive our official Honorable Discharge papers! What a *glorious* moment that was, to realize I was again a real life *CIVILIAN*!

(Home at Last! Hurrah!)

After congratulations and hugs and kisses from our loved ones, we all went our separate ways with families, and my family took me to a restaurant, for a celebration dinner. Then home! It was a time of great relief, arriving back in my hometown, whole and hearty, and in peak of physical condition, uninjured. Kept safe by my ever present Lord and Master Jesus Christ! That evening as I walked down the sidewalk to the soda fountain in town, it was dark; but I had no fear of a Jap sneaking up on me and stabbing me in the back. I also knew that in the unlikely situation that a thief was to approach me, I could kill him with one Judo blow. The release from tension was next to impossible to describe.

One left-over emotion was difficult, and required time to remove, and that was the hatred of the Japanese; instilled in us by our training and what we had been through for years. Yet I realized that hatred must be released if I was to grow spiritually, so I took that emotion to the Lord in prayer, confessing it, and asking for God's help to rid me of it. The burden was suddenly released, and I am no longer shackled by such hatred.

Another emotion that remains, is the immense patriotism, and idealism of the old American way, freedom. Every time I view a TV war movie and see the courageous sacrifice young men made to provide Americans with this freedom, and survival, it brings a welling up of emotion. I relate them to the men I served with or knew so well who did not return home, but gave their young lives so that we might live ours in our beloved land. They rate the greatest honor!

(A Civilian Again!)

It was time to get out of my Navy uniform, and into some civilian clothes, but new clothes were not available. My old clothes no longer fit me, and with great encouragement from my sister I determined to be one of the best dressed young men in town. This was to be a new beginning, old things were passed away, and all things were now new! With a Dad who had lots of pull with the clothiers in nearby Pomona, we visited the clothing store, and the manager fitted me with one of the rare outfits available. Then, he fitted me with a second set, which were usually only available to the rich and famous. Clothing was still in very short supply, and clothing manufacturers were switching their production from military uniforms to civilian clothing. I was fortunate, to have two sets of new clothes!

(At Home. Just Relaxing!)

I had determined to just relax, and do nothing for a few weeks, especially since we received our mustering out pay of three months wages. But that gets dull very fast, for an ex-Seabee, who continues the spirit of "Can Do." I visited my old employer, NTL, and was welcomed with open arms, and requested to come back to work in my old job. But I told them my plans were to go to college, and until that began, I would be occupied with other things. They did insist that if I wanted a job while in school, to come back and they would work out my hours.

Dr. Beckman said he would recommend me to the faculty at California Institute of Technology to enable me to take the entrance exams. So I bought textbooks of all the technical subjects and set about studying hard for the exams. I had been offered a job with brother Marius, up in the Sierras, working on his sawmill, so that is what I planned to do until I could make the trip back to Wisconsin to meet Cora. I also got word of the passing of the G.I. Bill which would provide supplemental expenses for college, and would pay tuition, and books.

My relationship with sweetheart Cora grew ever closer, as we corresponded, and as we made plans for our meeting. She was so pleased to hear of my plans for college, and my work ethics. Her family always believed in working hard and providing for their own, so my plans to work to save money for my education was well received by her family. It was indeed a strain to wait for our meeting from February until July, but we each were mature enough to realize we had to do things decently and in order. Our patience was not always that great though, and we were very impatient to meet!

(Cora Waiting for her Sweetheart)

My dad was a member of the San Dimas Chamber of Commerce, so I was invited by their program chairman to be a special speaker at their monthly dinner. I was to speak on what I saw as opportunities for a returned enlisted man for his future. Another man, an ex-Navy Officer was to speak on opportunities he saw for his own future. As I arrived at the dinner, I was led to a seat at the speaker's table, and introduced to the young returned Navy Officer, whose name was Richard M. Nixon. I didn't know him from Adam, but understood he was expecting to run for the seat as District Representative for Congress. History records the results of his election.

My good Christian Seabee friend Horace Trigg was invited by Dad to preach at our church, and I was to give my testimony, and introduce him to the congregation. He had been our Company Standard Bearer, so he very appropriately spoke on carrying the Banner for our Lord Jesus Christ. We had a great time together talking over events of our past, and looking to the future. He was to become a Missionary to Africa, and later a Minister in the States, after his wife died of cancer.

My plans were fulfilled, and the rest of my life will have to be told in memoirs to follow. The experiences God led me through in WWII was a great preparation for the rest of my life, and it provided another step in preparation for my entire career. One lesson I've learned is that life is composed of numerous steps, and we are prepared by our experiences to climb the stairway of life, step by step. Each step prepares the way for the next. We don't do it all in one giant leap, or we fall in the attempt!

I hope you will enjoy life, at least as much as I've enjoyed mine, and then you will be bound to have a great time ahead of you! My faith and trust in Jesus Christ brought me safely through the military experience of WWII, and I look forward to Him seeing me through each step of my future. I encourage you to do the same.

Clayton F. Rasmussen
—A VET

Stepping Stones of My Life:

February 16, 1946	Honorable Discharge from the USNR.
March 1, 1946	Started work at Rasmussen Lumber Mill.
July 1, 1946	Met Cora Jacobson after 2 1/2 years Correspondence.
July 12, 1946	Engaged to be Married to Cora, truly in Love.
September 7, 1946	Entered Mt. San Antonio College, in California.
June 6, 1947	Completed Freshman year of College.
June 15, 1947	Moved to Racine, Wisconsin, working one year.
June 25, 1948	Married to Cora, moved back to California for 2nd year of College.
June 7, 1949	Received Assoc. of Arts degree from Mt. San Antonio College.
September 10, 1949	Entered Univ. of Wisconsin, at Madison.
February 2, 1952	Received BS Degree in EE, with Physics Option.
February 15, 1952	Started work for Beckman Instruments as Engineer.
April 6, 1953	Daughter Cheryl born in Alhambra, California.
November 16, 1956	Son Gary born in Fullerton, California
July 1, 1959	Resigned from Beckman; Founded New Products Development Labs.

October 1, 1959	Merged with D.A.T.A. International, (Missionary Technical Support).
January 20, 1960	Started part-time with D.A.T.A., full-time with GTE Lenkurt, in Telecom Dev. Engineering.
January 14, 1964	Son David born in Fremont, California.
July 7, 1972	Resigned from Lenkurt, to build Bear Valley Cabin, at Ski Resort.
December 3, 1972	Became Associate Pastor/Business Manager at Calvary Baptist Church, Fremont, California.
March 30, 1975	Resigned from Church & School, started part-time.
March 15, 1977	Started work at Hughes Aircraft Co. as Systems Engineer/Project Manager
May 31, 1988	Retired from Hughes, to care for Cora's health.
June 15, 1989	Cora died from Cancer, after our 41 years of wonderful Marriage.
February 8, 1992	Hughes Representative at Pentagon on EPLRS Field Demonstration. (Consultant). 2-years.
July 9, 1992	Week Missions Trip, to Island of Dominica, Preaching, Teaching & Child Evangelism.
March 3, 1993	Bought my first Personal Computer Started Memoirs.
May 5, 1998	Sold my Rancho Cucamonga House. Moved in with son Gary's Family in Roseville.
Until the Rapture!	Continued Writing Memoirs and Biographies, Enjoying Family, Bible Teaching, Traveling, Flying, Research, Serving the Lord.

Benefits of the G.I. Bill, & a Working Wife

(Entering Mt. San Antonio College—1946)

(Graduate, BS in EE w/Physics Option, Leaving Halls Of Ivy—1952)

Addendum I
About the Author

Clayton F. Rasmussen was born in 1922, in a rural community in Michigan. His father was a Baptist minister who later served in churches in Wisconsin, Illinois, Pennsylvania, and California. Clayton was the youngest of five children, moving to each new location with his parents.

He received most of his Elementary and High School education in Pennsylvania. Upon graduation he preceded his parents to California in search of employment in order to finance a college education. However, college would have to be deferred a few years. First employed by National Technical Laboratories, a company founded and managed by Dr. A. O. Beckman, he was trained, and became an experimental machinist in the Engineering Laboratory.

Deferred from the draft when WWII broke out, he continued work at NTL until 1943 when he entered the Navy Seabees as a machinist. Serving in the Pacific theatre till after the war, he returned to civilian life and to college under the new GI Bill.

Married to his Wisconsin Sweetheart after his first year of college, he graduated from the University of Wisconsin as an Electrical Engineer, with a combined major in Physics. He was re-employed in 1952 by Dr. Beckman at the Beckman Instruments Corporation (renamed from NTL). There he became Systems Project Engineer/Manager developing Data Processing Systems for the Petroleum industry.

In late 1959, he left Beckman's employ, to found New Product Development Lab, which later merged with D.A.T.A. International. They developed special technical products and provided solutions to technical problems of Missionaries in remote areas of the world. This fledgling company eventually became the "People to People" program under President Eisenhower, and then President Kennedy took the concept, and politicized it and called it the "Peace Corps." The author had left the endeavor as it became a political program.

Employed as an Engineer by GTE Lenkurt he was instrumental in the development of Telecommunications Systems. The most significant of these was the new miniaturized 36A System, utilized by industry for secure communications. The most famous applications were by the Nixon White House, and the Nation of Israel.

Returning to Southern California after his wife developed cancer, he was employed by Hughes Aircraft working as Systems Project Manager on Defense projects: Radar, Sonar, and Smart weapons guidance systems.

Upon retirement he cared for his wife until her death in 1989. Then began a part-time consulting work, again at Hughes, representing them at the Pentagon for field demonstrations of a new battlefield communications system, and extended field training of the Military in its use. His primary activities at present are in further writing projects.

Addendum II

Special Features of WWII "Special Assignment" Book

1. It covers the *entire military career* of a wartime enlisted man in the Navy Seabees, from civilian-to-military-to-civilian life again.
2. It provides *chronologically* the events experienced in this military period and specialty force.
3. To my knowledge, it is the first personal account describing the *unique training* of a Seabee, and an enlisted man's experience through it.
4. It displays the *effective use of censorship* at a time of critical need to maintain secrecy from the enemy, and to uphold morale of the citizenry at home. It prevented political and subversive interference with the military in carrying out their plans and strategy.
5. It provides *historical accuracy* for a period of time and through events that could be otherwise misinterpreted by later revisionist historians.
6. It provides details never before revealed about the *first nuclear events,* relating to the A-bomb's successful deployment.

7. It testifies to the *spiritual resource available* to a young Christian under stressful, traumatic events, and the spiritual growth resulting.
8. It shows the *personal relationships* that can exist between officers and enlisted men when each respects the other's responsibility.
9. It provides evidence of the *unique individualism* and interests that continue through all walks of life.
10. It shows to young people that a person should "*be their own unique self*" at all times, and have the *right attitude* and take full *responsibility* for their required duties.
11. My picture, "*The Dawn of Peace,*" has never before been published and is "*one of its kind.*" It was taken at dawn on the very day Japan surrendered, to end World War Two.
12. It displays the principle that *life is a series of growth steps*; each step leading to the next.
13. It shows that even *difficult circumstances* can be enjoyed, when a person endeavors to make the best of it, looking for continued growth and maturity. This is especially *true of a Christian,* trusting in Jesus Christ and His Word, for leadership.
14. It reveals the *dramatic changes* that have occurred *in a single generation*; in technology, methods, and World view.
15. Most of the *pictures* used as illustrations were personally photographed by the author. Sketches were creatively designed by the Battalion artist, Bill Valentine.

Addendum III

Utilization of "Special Assignment" as a WWII Teaching Resource.

1. The events recounted are historically accurate, and the timing of each is precise.
2. These accounts provide emotions, as well as historical events.
3. The book is written from the viewpoint of one man living through those events, not simply from a distant historical perspective. So it capitalizes on things of the author's personal interests.
4. The troops were mostly unaware of the planned military strategy, they only did what was evidently required of them each day. The desired end result of a military maneuver was not necessarily evident, requiring obedience to authoritative commands.
5. Under combat conditions, the Seabees were not required to salute, or have formal dress as they did in the States. Relationships of officers and men were similar to occupational relationships of civilians.
6. Language used in this book is not always "politically correct," by the World's standards today.

7. Events and information concerning my friend, Dr. A. O Beckman were confirmed by him as we met for a visit in July, 1997.
8. Information and experiences related to the Hiroshima Bomb, was recently reviewed for confirmation, with two members of the "Enola Gay" crew, Navigator Van Kirk, and Radioman Nelson.
9. The information contained in the Addendum to the Atomic Bomb chapter needs to be specially emphasized. The dropping of warning leaflets before the bombing is virtually never discussed in history books.
10. Bringing the millions of troops home after WWII was a monumental task requiring several months. Ships were the primary means used. Compared with today's method by air, this was a tremendous change.
11. The Navy Seabees, as a military entity, was originated during WWII, just one year before the author entered the Service. They spawned the UDT (Frogmen), which in later years became the Navy Seals.
12. This book has been used in class study especially for Jr. High and Sr. High Schools. However it has proven useful as a teaching tool down to Grade 3. One 5th Grade girl commented: "I never dreamed I would ever know so much about WWII, as I learned in this class." The book has proved very useful in school libraries, as an added resource. In relation to the "Manhattan Project," and work at NTL, it has been used as a resource in the biography of the world renowned Scientist, Dr. A. O. Beckman.—Clayton F. Rasmussen

Addendum IV

Meet the People

Machine Shop Associates

P. Derflinger, V. Ranke, H. Brunetti L. Tressler C. Rasmussen, C. Cooper

Chief H. Ward M. Rakowicz Warrant Officer White

This Was the Entire Machine Shop Crew, and Officers on Eniwetok.
These were Clay's Daily Associates in Machine Shop Duties.

Machine Shop & Crew on Tinian.
The Crew Expanded as Needed to Support the Larger Workload.
The Battalion Had a Reserve of Experienced Machinists.

Our 6-man Tent on Tinian

Our Tentmates

Kennedy & Ganski

Walters

Trigg & Rasmussen

Smith

LaVergne, TN USA
05 August 2010
192184LV00004B/303/A

9 781414 100043